T0171664

COMPRESSED AIR PROPULSION SYSTEM
TO POWER THE CAR OF THE FUTURE

ROBERT J. RICHEY

authorHOUSE®

AuthorHouse™
1663 Liberty Drive
Bloomington, IN 47403
www.authorhouse.com
Phone: 1 (800) 839-8640

Published by AuthorHouse 02/23/2016

ISBN: 978-1-5049-8130-9 (sc)
ISBN: 978-1-5049-8129-3 (e)

Print information available on the last page.

Any people depicted in stock imagery provided by Thinkstock are models, and such images are being used for illustrative purposes only. Certain stock imagery © Thinkstock.

This book is printed on acid-free paper.

INDEX

INTRODUCTION

The world community today is faced with several dilemmas. The first of these is overpopulation. This problem is contributing to hunger and even starvation in parts of the third world. A second dilemma is pollution of the atmosphere and the waters of the world. This pollution is changing the world's atmosphere and causing global warming on a grand scale. The pollution of the continent waterways endangers human and animal health. Contamination of the planets seas and lakes endangers the survival of the fish and other creatures that dwell in the seas.

Some steps are being taken but not on a scale to address these problems that would lead to a final solution.

At least 86% of the world's energy is produced using fossil fuels. Hydroelectric sources produce about 6.5% while Nuclear produces about 8.5%. Unfortunately disposal of nuclear waste is a continuing problem. The life of nuclear waster stretches into the infinite. Production of energy from a variety of sources such as geothermal, solar and burning of waste only produces about 9%.

One of the most troublesome problems is a result of the increasing congestion on the nation's highways. Millions of vehicles sit idly in traffic for hours on end expelling millions of tons of carbon dioxide into the atmosphere. The drivers of these vehicles and their passengers are engulfed in a river of monoxide on a daily basis. The potential for health issues in the future are catastrophic.

There is no solution to the production of the world's increasing need for electricity to supplement or replace fossil fuels in view today. Although the availability of fossil fuels is vast it still is limited.

Although there is no solution to the problems itemized here there are steps being taken in some measure to reduce the adverse effects of the mandatory use of fossil fuels. The installation of solar arrays is being pursued on an increasing scale.

The proposed invention on the pages of this book offers a new and different way of propelling vehicles on the nation's highway. It has the advantage of the electric in that it only operates when the vehicle is in motion. It also has the feature of being pollution free during operation. The air under pressure is merely expelled back into the atmosphere. The source of the compressed air is removed from the nations roadways and transferred to a large number of filling stations of the future located at convenient distances from each other or along the shoulders of the nation's various roadways.

Use of alternate means of propelling vehicles like the Car for the Future will save a lot of fossil fuels and at the same time reduce the amount of carbon dioxide and monoxide expelled into the atmosphere.

United States Patent	9,217,329
Richey	December 22, 2015

Compressed air propulsion system

Abstract

A compressed air propulsion system supplies air utilized by a pair of opposing cylinders and their associated pistons and push/pull rods to cause a pair of sprockets to rotate clockwise in a controlled manner. The two pistons are acted upon by the cycling of various valves which introduces and/or vents compressed air as directed by a computer using a downloaded program through wireless interfaces. One of the two sprockets in turn, through additional sprockets/chain/axle devices, is utilized to supply drive torque to a vehicle transmission. The other axle which is connected to the remaining sprocket and through an additional chain/sprocket/axle device operates a direct current generator which produces electricity for charging a battery.

Inventors:	Richey; Robert John (Campbell, CA)			
Applicant:	Name	City	State	Country Type
	Richey; Robert John	Campbell	CA	US
Family ID:	54847915			
Appl. No.:	14/756,548			
Filed:	September 16, 2015			

Related U.S. Patent Documents

Application Number	Filing Date	Patent Number	Issue Date
14256754	Apr 18, 2014	\<TD\< TD\>	\<TD\< TD\>

1

Current U.S. Class:	**1/1**
Current CPC Class:	F01B 11/001 (20130101); F01B 7/14 (20130101); F01B 25/04 (20130101); F01B 29/10 (20130101); F01L 31/16 (20130101)
Current International Class:	F16D 31/02 (20060101); F01B 7/14 (20060101); F01B 11/00 (20060101); F01B 25/04 (20060101); F01B 29/10 (20060101); F01L 31/16 (20060101)
Field of Search:	;60/370,408,409

References Cited [Referenced By]

U.S. Patent Documents

3485221	December 1969	Feeback
3980152	September 1976	Manor
4370857	February 1983	Miller
4560323	December 1985	Orchard
5515675	May 1996	Bindschatel
6629573	October 2003	Perry
7028482	April 2006	Cho
7315089	January 2008	Lambertson

Primary Examiner: Leslie; Michael
Attorney, Agent or Firm: Hill; Robert Charles

Parent Case Text

CROSS-REFERENCE TO RELATED APPLICATIONS

This application is a Continuation-In-Part of U.S. Ser. No. 14/256,754 filed Apr. 18, 2014 for THE CAR OF THE FUTURE POWERED BY COMPRESSED AIR.

Claims

The invention claimed is:

1. A compressed air propulsion system, comprising: a compressed air tank; first and second identical opposing cylinders each including a piston connected to a piston rod, each piston dividing the cylinder into two chambers; each chamber connected to inlet valves and outlet valves; the cylinder chambers connected to the compressed air tank via the inlet valves; first and second sprockets joined by a common chain, the sprockets mounted on corresponding first and second axles; the piston rods of the first and second cylinders are connected to a corresponding one of the first and second sprockets at a pivot points; each axle having a fixed disc mounted adjacent to the sprocket, wherein the fixed disc includes plural light beam emitter/receptor devices that send signals to an electronic control unit for controlling the opening and closing of the inlet valves and outlet vales based on the position of interrupters mounted to each sprocket; the first axle connected to the first sprocket supplying drive torque to a vehicle transmission; and the second axle connected to the second sprocket supplying drive torque to an electronic generator which charges a battery.
2. The compressed air propulsion system of claim 1 wherein: each cylinder has 2 inlet valves, 2 outlet valves, and a pressure gauge.
3. The compressed air propulsion system of claim 1 wherein: each emitter/receptor device has a light beam interrupter attached to the back side of a sprocket.
4. The compressed air propulsion system of claim 3 wherein: a bracket attached to a fixed disc which houses a light beam emitter, and a light beam receptor having a light beam therebetween.
5. The compressed air propulsion system of claim 4 wherein: rotation of the sprocket causes a finger on the light beam interrupter to break the light beam between the light beam emitter and the light beam receptor.

Description

BACKGROUND OF THE INVENTION

1. Field of the Invention

The present invention relates to a propulsion system, and in particular to a propulsion system using compressed air in which supplies drive torque to a vehicle transmission as well as drive torque to an electronic generator for charging a battery.

2. Description of the Related Art

The proposed invention can be related to providing propulsion of a vehicle in a more efficient and less expensive manner, meanwhile reducing the amount of pollution being released into the atmosphere. The current method is not only wasteful and inefficient but is prone to producing vast amount of pollutants into the atmosphere on a daily basis while not doing useful work at the same time. This vast amount of pollution is believed to be contributing to the global warming of the earth and all of the ills that are attendant with it, like flooding of lower coastal regions etc. The basic reason, as concerns motor vehicles, is the ever increasing congestion on all of the nations roadways whether city streets; state highways or federal freeways. This massive congestion results in prolonged delay of vehicles of all types in trying to reach their destinations. The impressive gas mileages that all of the automobile manufacturing companies extol are meaningless when all of the vehicles during the rush hour spend long periods of time stalled and idling in traffic. Although the electric car is subject to the same traffic delays as all of the other vehicles propelled by a different method such as gas or diesel engine, still the electric car is in the off mode when not in motion. Yet the problem with the electric car and hybrid car, even though to a lesser degree, is the limited amount of storage of electricity available with the current technology in battery manufacturing.

BRIEF SUMMARY OF THE INVENTION

The compressed air propulsion system of the present invention includes a compressed air tank. First and second identical opposing cylinders have a piston connected to a piston rod. Each piston divides the cylinder into two chambers and each chamber is connected to inlet and outlet valves with the inlet valves connected to the compressed air tank. First and second sprockets joined by a common chain are mounted on corresponding first and second axles. The piston rods of the first and second cylinders are connected to corresponding first and second sprockets at a pivot point. Each axle has a fixed disc mounted adjacent to the sprocket. The fixed disc includes plural light beam emitter/receptor devices that send signals to an electronic control unit for controlling the opening and closing of the inlet and outlet valves based on the position of interrupters mounted to each sprocket. The first axle connected to the first sprocket supplies drive torque to a vehicle transmission and the second axle connected to the second sprocket supplies drive torque to an electronic generator which charges a battery.

BRIEF DESCRIPTION OF THE DRAWINGS

FIG. 1 is an end view of the present invention.

FIG. 2 is a block diagram of the present invention.

FIG. 3 is a top view of the present invention.

FIG. 4 is a view taken along line A-A of FIG. 3.

FIG. 5 is a view taken along line B-B of FIG. 3.

FIG. 6 is a top view of the drive shaft and its housing.

FIG. 7 shows five brackets affixed to fixed disc 502.

FIG. 8 shows 4 brackets affixed to fixed disc 604.

FIG. 9 is a listing that depicts all of the various modes of the two cylinders for one cycle of the sprockets.

DETAILED DESCRIPTION OF THE INVENTION

Many of the details of the present invention are shown in FIGS. 1 through 5 including a compressed air tank 436 (FIGS. 2 and 3). First 202 and second 214 identical opposing cylinders include a piston connected to a piston rod. First cylinder 202 has piston 204 connected to piston rod 206 while second cylinder 214 has piston 216 connected to piston rod 218.

Each piston is divided into two chambers. As shown in FIG. 1 first cylinder 202 has chambers 231 and 233 while second cylinder 214 has chambers 235 and 237.

Each chamber has inlet valves and outlet or vent valves. Left chamber 231 of first cylinder 202 is provided with inlet valve 230 and vent or outlet valve 232 and pressure gauge 246. Right chamber 233 of first cylinder 202 is provided with inlet valve 236 and vent or outlet valve 234 and pressure gauge 248. Likewise left chamber 235 of second cylinder 214 is provided with inlet valve 238 and vent or outlet valve 240 and pressure gauge 252. Right chamber 237 of second cylinder 214 is provided with inlet valve 244 and vent or outlet valve 242 and pressure gauge 250.

The cylinder chambers are connected to the compressed air tank 436 via inlet valves. As shown in FIG. 3 inlet line 431 connects tank 436 with first cylinder 202 via inlet valves 230 and 236 while inlet line 433 connects tank 436 with second cylinder 214 via inlet valves 238 and 244.

First sprocket 226 and second sprocket 228 are joined by common chain 229. First sprocket 226 is mounted on first axle 241 and second sprocket 228 is mounted on second axle 243.

The piston rods of the first and second cylinders are connected to a corresponding one of the first and second sprockets at a pivot point. As shown in FIG. 1, piston rod 206 of first cylinder 202 is connected to first sprocket 226 at pivot point 212 via joint 208 and push rod 210 while piston rod 218 is connected to second sprocket 228 at pivot point 224 via joint 220 and push rod 222. Since pivot 212 is at

90.degree. and pivot point 224 is at +1-180.degree. the pivot points are 90.degree. out of phase.

Each axle has a fixed disc mounted adjacent to the sprocket. As shown in FIG. 4, first axle 241 has fixed disc 502 mounted adjacent to first sprocket 226. In FIG. 5 second axle 243 has a fixed disc 604 mounted adjacent to second sprocket 228.

Each fixed disc 502 and 604 includes plural light beam emitter/receptor devices 530 (FIG. 4) and 630 (FIG. 5) that send signals to an electronic control unit 435 for controlling the opening and closing of the inlet valves 230 236, 238, 244 and outlet valves 232, 234, 240, 242 based on the position of interrupts 518 and 602 mounted to each sprocket 226 and 228.

The first axle 241 is connected to the first sprocket 226 supplying drive torque to a vehicle transmission (FIG. 3); and the second axle 243 is connected to the second sprocket 228 supplying drive torque to an electronic generator (FIG. 3) which charges a battery.

As shown in FIG. 3 the electronic control unit 435 includes a computer 430, a master wireless interface device 430.1 and a monitor 432. Various wireless interface devices (WID) serve as the wireless link with the master wireless interface device 430.1 which enables the computer 430 to command the opening and closing of inlet and/or outlet valves in a preprogrammed order.

FIG. 1 shows first cylinder 202 having WID 230.1 associated with inlet valve 230, WID 232.1 associated with outlet valve 232, WID 234.1 associated with outlet valve 234, WID 236.1 associated with inlet valve 236, WID 246.1 associated pressure gauge 246, and WID 248.1 associated with pressure gauge 248. Likewise, second cylinder 214 had WID 238.1 associated with inlet valve 238, WID 240.1 associated with outlet valve 240, WID 242.1 associated with outlet valve 242, WID 244.1 associated with outlet valve 244, WID 250.1 associated with pressure gauge 250, and WID 252.1 associated with pressure gauge 252.

The master wireless interface device 430.1 is the link between the computer 430 and the various switches, pressure gauges and emitter/receptor devices.

The opening and closing of the various switches are dependent on which light beam has been intercepted or blocked by the light beam interrupter. For cylinder 202 that would be 518 from FIG. 4 and for cylinder 214 that would be light beam interrupter 602 of FIG. 5.

Each light beam that is blocked informs the computer to command a particular inlet valve to change to the open or closed mode or for a particular vent valve to change to the open or closed mode.

The changes by valves of both cylinders are made at the same time. The only difference is that each sprocket has its own light beam interrupter attached to the back of it. And since the two cylinders work together it is required that the particular valves open and close in conformance with instructions that have been programmed in the computer program that has been down loaded on the computer hard drive.

The device is intended to work when each piston produces power in sequence, not simultaneously. When one piston is in the push or pull power mode the other piston is in the vent mode.

As stated, the device operates in several modes. First is the start up mode in the clockwise direction, then operation mode. This can be of the push or pull mode by either of the pistons. There is also the shut down mode where the device is brought to a halt. These modes apply only to the two cylinders.

In FIG. 2 compressed air from tank 436 goes through inlet line 431 to first cylinder 202 to drive push rod 210 which rotates first sprocket 226 and downstream sprocket 506 on first axle 241. Then downstream sprocket 506 by means of chain 438 rotates connecting sprocket 722 which connects to a vehicle drive shaft. In the same manner compressed air from tank 436 goes through inlet line 433 to second cylinder 214 to drive push rod 222 which rotates second sprocket 228 and downstream socket 608 on second axle 243. Then downstream socket 608 by means of chain 440 rotates connecting sprocket 434 which connects to electric generator 424. The electric generator supplies power to battery pack 422.

FIG. 3 has another mode that involves the valves and pressure gauges that are involved in introducing air under high pressure into the various tanks from one or more external sources. One would be from a filling station of the future or from a compressor located in the owners garage.

Electric motor 418 operates air compressor 414 which supplies compressed air via line 417 to tank 436 through valve 412. Also, station 420 supplies compressed air via line 403 to tank 436 through valve 402 when valve 406 is closed. Pressure within tank 436 should be around 600 psi while the pressure within cylinders 202 and 214 as regulated by the various pressure gauges is between 10-15 psi.

FIG. 3 also shows connecting sprocket 434 rotating generator axle to supply electricity to DC generator 424 which goes into battery 422 via line 423. DC from the battery 422 goes through inverters 426 and 428 to become AC which then goes through electric lines 429 to supply power to all valves, WIDS, pressure gauges and other equipment that require electricity.

The lower portion of FIG. 4 is a larger view of the emitter/receptor assembly, generally indicated 530, which has a light beam interrupter 518 attached to the back side of first sprocket 226 and bracket 531 attached to fixed disc 502. Bracket 531 houses light beam emitter 526 and light beam receptor 532 having light beam 533 therebetween. As sprocket 226 is rotated finger 519 of light beam interrupter 518 breaks the light beam 533 between emitter 526 and receptor 532, the consequences of which will be described later.

Supports 504 and 508 on base 524 house first axle 241 which is connected to first sprocket 226 and downstream sprocket 506. Flange 514 attaches to support 504 and flange 510 attaches to support 508. Fixed disc 502 is connected to support 504 by bolts 520 and 522. Brake 534 is on axle 241 and is activated by WID 534.1.

The lower portion of FIG. 5 is a larger view of the emitter/ receptor assembly, generally indicated 630, which has a light beam interrupter 602 attached to the back side of second sprocket 228 and a bracket 631 attached to fixed disc 604. Bracket 631 houses light beam emitter 632 and light beam receptor 634 having light beam 636 therebetween. As sprocket 228 is rotated finger 603 of light beam interrupter 602 breaks the light beam 636 between emitter 632 and receptor 634, the consequences of which will be described later.

Supports 606 and 610 on base 628 house second axle 243 which is connected to second sprocket 228 and downstream sprocket 608. Flange 626 attaches to support 606 and flange 624 attaches to support 610. Fixed disc 604 is connected to support 606 by bolts 614 and 616.

FIG. 6 is an enlarged view of a portion of FIG. 3 wherein supports 720 and 724 hold housing/drive shaft 718 which engages differential 708. Rotation of sprocket 722 rotates drive shaft 718 which through differential 708 rotates left axle 710 and right axle 716 thereby rotating wheels 702 and 704. Brake 706 is adjacent wheel 702 while brake 712 is adjacent wheel 704.

FIG. 7 shows brackets 810, 812, 814 and 816 attached to fixed disc 502 at angles of +45, +135, -135, and -45 respectfully. Bracket 818 at +90 is de-energized in operational mode and is activated during the shut down mode. As described above for FIG. 4 each bracket houses a light beam emitter and a light beam receptor having a light beam therebetween. WIDs 810.1, 812.1, 814.1, 816.1, and 818.1 are associated with their respective brackets.

FIG. 8 shows the four brackets on fixed disc 604. Bracket 902 is attached at +45.degree., bracket 904 at 45.degree., bracket 906 at +135.degree., and bracket 908 at -135.degree.. WIDs 902.1, 904.1, 906.1 and 908.1 are associated with their respective brackets. As described above for FIG. 5 each bracket houses a light beam emitter and a light beam receptor having a light beam therebetween.

FIG. 9 lists the open and closed positions of the inlet and vent valves as the sprockets 226 and 228 complete one operational cycle. This will be explained in detail later on.

The invention has two power units consisting of two cylinders that are in opposition. Each cylinder has an enclosed piston which divides the cylinder into two chambers. Each chamber of each cylinder is furnished with an inlet valve and an outlet or vent valve.

Each inlet valve allows air under pressure from the internal storage tank(s) to flow into the particular chamber in conformance with instructions previously included in a computer program downloaded on the computer hard drive. This pressured air in turn applies force to a piston. Each piston is connected to a push/pull bar that moves right or left with the piston. Each horizontal push/pull bar in turn is connected to a second push/pull bar that pivots at the connection point with its respective horizontal push/pull bar.

Functioning of the individual vent valves is identical to that of any inlet valves. Commands to change from the open or closed mode are transmitted wirelessly through commands from the electronic control unit through electronic pulses.

Any one of the four individual pressure gauges serves only to provide a constant readout to the computer wirelessly of the pressure existing in a particular chamber at a particular moment in time. Pressure gauges play no active role in generating force by the power unit.

The two sprockets rotate in unison due to a common chain that links them together. Each of the two sprockets is identical in size and shape and functions the same way.

Operation of the inlet valves or vent valves in either the open or closed position is determined by instructions from and through the master wireless interface device 430.1 to the individual wireless interface devices that are part of each of the inlet or vent valve assemblies.

The compressed air propulsion system operates in one of three modes. These modes are startup, operation, and shutdown.

Each of the two cylinders operates in one of four modes. The modes are push, vent, pull and vent. These modes are repetitive.

When either cylinder is in the push or pull mode the opposite cylinder is in the vent mode.

Only one inlet valve of the four inlet valves on the cylinders can be in the open position mode at any time during the operation mode. More than one vent valve in either cylinder may be in the open mode at any one time.

Startup Mode

When the system is placed in the startup mode on computer command inlet valve 230 is ordered to the open mode. Pivot point 212 rotates between locations +90 and +135 relative to fixed disk 502. This causes finger 519 of light beam interrupter 518 attached to the back side of sprocket 226 to break light beam 533 housed within bracket 812 at location +135 of fixed disk 502 resulting in inlet valve 230 changing from the open mode to the closed mode.

Pivot point 224 has been resting at location +180 relative to fixed disk 604 during the shutdown mode. Pivot point 224 rotates from +180 to location -135. This causes finger 603 of light beam interrupter 602 attached to the back side of sprocket 228 to break the light beam 636 housed within bracket 908 of fixed disk 604 causing inlet valve 244 to change from the off mode to the on mode. Meanwhile vent valve 242 changes from the on mode to the off.

The power unit of the invention continues to operate on commands transmitted by wireless pulses to the various valves and switches until the next shutdown operation is reached.

Operation Mode

For cylinder 202 the push mode occurs when inlet valve 230 is in the open mode, vent valve 232 is in the closed mode and inlet valve 236 is in the closed mode while vent valve 234 is in the open mode. At the same time inlet valve 244 of cylinder 214 is in the closed mode and vent valve 242 in the right chamber of cylinder 214 is in the open mode. Inlet valve 238 is in the closed mode while vent 240 is in the open mode. Air flows into the left chamber of cylinder 202 and out through vent valve 234 of the right chamber. During the same interval in time air flows in or out of either vent valve 240 or 242 as the two sprockets rotate.

Cylinder 202 is in either of the two vent modes when both inlet valves 230 and 236 are in the closed mode and the two vent valves 232 and 234 are in the open mode. During the same interval in time inlet valves 238 and 244 of cylinder 214 are in the closed mode. Conversely vent valves 240 and 242 are in the open mode. Air may flow in or out of either chamber depending on the direction the particular piston is moving.

Cylinder 202 is in the pull mode when pressure enters chamber 233 through inlet valve 236 which is in the open mode and vent valve 234 located in the same chamber is in the closed mode. Inlet valve 230 is in the closed mode and vent valve 232 is in the open mode. For cylinder 214 both inlet valves 238 and 244 are in the closed mode and vent valves 240 and 242 are in the fully open mode.

Cylinder 214 is in the push mode when inlet valve 244 is in the open mode, vent valve 242 is in the closed mode, inlet valve 238 is in the closed mode and vent valve 240 is in the open mode. At the same time both inlet valves 230 and 236 of cylinder 202 are in the closed mode while both vent valves 232 and 234 are in the fully open mode.

Cylinder 214 is in either of the two vent modes when inlet valves 238 and 244 are in the closed mode and vent valves 240 and 242 are in the fully open mode. Cylinder 202 is in the either the push mode or the pull mode.

Cylinder 214 is in the pull mode when inlet valve 238 is in the open mode, vent valve 240 is in the closed mode, inlet valve 244 is in the closed mode and vent 242 is in the open mode.

As shown in FIG. 9, cylinder 202 is also in the push mode when pivot point 212 rotates between +45 and +135 relative to fixed disk 502 (FIG. 3 & FIG. 7) and in the vent mode when pivot point 212 rotates between +135 and -135. It is in the pull mode when pivot point 212 rotates between -135 and -45. It is in a second vent mode when pivot point 212 rotates between -45 and +45. Any following rotation repeats itself. The rotation of either sprocket depends on direction of rotation of either push/pull bar depending on which is in the power mode.

Cylinder 214 is in the push mode when pivot point 224 rotates between -135 and -45 of fixed disk 502. It is in one of two vent modes when pivot point 224 rotates between -45 and +45. It is in the pull mode when pivot point 224 rotates between +45 and +135. It is in a second vent mode when pivot point 224 rotates between +135 and -135.

Opening or closing of any of the four inlet valves of the two cylinders or the vent valves thereof depends on pulses transmitted through wireless interfaces devices (WIDs). The pulses generated by the master wireless interface device 430.1 are determined by instructions embedded in the computer program. These pulses are transmitted in a predetermined order to achieve a smooth rotation of the two sprockets. Rotation of the sprockets in turn causes the axles they are mounted on to rotate. One axle 241 serves to provide motion for a vehicle while rotation of a second axle provides rotation of the armature of the DC generator to generate electricity to power the invention.

In operation starting with cylinder 202 in the push mode with inlet valve 230 in the open mode, vent valve 232 in the closed mode, vent valve 234 in the open mode while inlet valve 236 is in the closed mode. In the same interval of time both inlet valve 238 and 244 of cylinder 214 are in the closed mode and vent valves 240 and 242 are in the open mode. Air under pressure is introduced through inlet valve 230 and flows out of right chamber vent valve 234. Air that is in the chambers 235 and 237 flows in or out of either chamber depending on the motion of piston 216.

Each sprocket has a light beam interrupter device (LBI) mounted on the reverse side of the sprocket. Sprocket 226 has LBI 518 (refer to FIG. 4) mounted along a radius extending from the center of axle 241 (refer to FIG. 1) vertically to a position opposite to pivot point 212. LBI 602 (refer to FIG. 5) is mounted along a radius extending from the center of axle 243 (refer to FIG. 1) to a point opposite to pivot point 224).

The finger 519 of LBI 518 breaks in succession the light beams 533 housed in brackets 818, 812, 814 and 816 (refer to FIG. 4 & FIG. 7).

Meanwhile the finger 603 of LBI 602 breaks successively through the light beams 636 housed in brackets 902, 906, 908 and 904 (refer to FIG. 5 & FIG. 8).

For the emitter/receptor assembly 530 mounted on the face of fixed disc 502 (refer to FIG. 3) pulses are generated when the finger 519 of LBI 518 momentarily breaks the light beam existing between the particular emitter/receiver device gap. This also occurs simultaneously when the finger 603 of LBI 602 momentarily breaks the light beam existing between the particular emitter/receptor devices that are mounted on the face of fixed disk 604.

The emitter/receptor housed in bracket 818 is provided with electrical power only during the shutdown mode but is not active in the startup or operational mode.

Blockage of the light beam within bracket 810 at location +45 on fixed disk 502 causes the computer through the master wireless interface device 430.1 to transmit a pulse or pulses that cause inlet valve 230 to change from the closed mode to the open mode. Another pulse generated at the same time causes vent valve 232 to change from the open mode to the closed mode as cylinder 202 changes from the vent mode to the push mode. Other pulses when LBI 602 breaks the light beam at -45 on fixed disk 604 (refer to FIG. 8) causing inlet valve 244 to change from the open mode to the closed mode and another pulse causes vent valve 242 to change from the closed mode to the open mode as cylinder 214 changes from the push mode to the vent mode.

Blockage of the particular light beam of any of the emitter/receptor devices in like manner causes inlet and outlet valves of both cylinders to command the particular inlet or vent valves to open or close as cylinders change from one mode to the next mode.

Shutdown Mode

When the shutdown activating device (not shown) is in the on mode and on the second rotation of pivot point 212 through location +45 of fixed disk 502 inlet valve 230 is changed to the open condition until pressure gauge 246 by pulse verifies that the design pressure of 10 psi in chamber 231 has been reached. At this time inlet valve 230 is placed in the closed mode. This is imposed on inlet valve 230 so that the pressure in chamber 231 decreases as sprocket 226 continues to rotate and chamber 231 increase in volume causing a decrease in pressure. This preprogrammed decrease in pressure results in a decrease in the force exerted on the face of piston 204. When LBI 518 breaks the light beam E/R/D within 818 the electric current to E/R/D within 818 is placed in the off mode. At the same instant brake 534 mounted on axle 241 is placed in the engaged mode and the rotation of axle 241 is brought to a halt. In the shutdown mode both cylinders rest in the vent mode where inlet valves 230 and 236 of cylinder 202 rest in the closed mode and vent valves 232 and 234 rest in the open mode. Inlet valves 238 and 244 of cylinder 214 rest in the closed mode while vent valves 240 and 242 rest in the open mode.

Although particular embodiments of the present invention have been described and illustrated, such is not intended to limit the invention. Modifications and changes will no doubt become apparent to those skilled in the art, and it is intended that the invention only be limited by the scope of the appended claims.

* * * * *

(12) **United States Patent**
Richey

(10) Patent No.: **US 9,217,329 B1**
(45) Date of Patent: **Dec. 22, 2015**

(54) **COMPRESSED AIR PROPULSION SYSTEM**

(71) Applicant: **Robert John Richey**, Campbell, CA (US)

(72) Inventor: **Robert John Richey**, Campbell, CA (US)

(*) Notice: Subject to any disclaimer, the term of this patent is extended or adjusted under 35 U.S.C. 154(b) by 0 days.

(21) Appl. No.: **14/756,548**

(22) Filed: **Sep. 16, 2015**

Related U.S. Application Data

(63) Continuation-in-part of application No. 14/256,754, filed on Apr. 18, 2014, now abandoned.

(51) **Int. Cl.**
F16D 31/02	(2006.01)
F01B 7/14	(2006.01)
F01B 11/00	(2006.01)
F01B 25/04	(2006.01)
F01B 29/10	(2006.01)
F01L 31/16	(2006.01)

(52) **U.S. Cl.**
CPC *F01B 7/14* (2013.01); *F01B 11/001* (2013.01); *F01B 25/04* (2013.01); *F01B 29/10* (2013.01); *F01L 31/16* (2013.01)

(58) **Field of Classification Search**
CPC F01B 7/14; F01B 11/001; F01B 25/04; F01B 25/14; F01B 29/10; F01L 31/16
USPC 60/370, 408, 409
See application file for complete search history.

(56) **References Cited**

U.S. PATENT DOCUMENTS

3,485,221	A	*	12/1969	Foebeck	F01D 7/14 123/51 AA
3,980,152	A	*	9/1976	Manor	B60G 13/14 180/165
4,370,857	A	*	2/1983	Miller	B60K 3/02 180/165
4,560,323	A		12/1985	Orchard	
5,515,675	A	*	5/1996	Bindschatel	F02B 69/06 60/370
6,629,573	B1	*	10/2003	Perry	F01B 17/02 180/165
7,028,482	B2	*	4/2006	Cho	B60K 1/00 60/407
7,315,089	B2	*	1/2008	Lamberson	B60L 11/005 180/302

* cited by examiner

Primary Examiner — Michael Leslie
(74) *Attorney, Agent, or Firm* — Robert Charles Hill

(57) **ABSTRACT**

A compressed air propulsion system supplies air utilized by a pair of opposing cylinders and their associated pistons and push/pull rods to cause a pair of sprockets to rotate clockwise in a controlled manner. The two pistons are acted upon by the cycling of various valves which introduces and/or vents compressed air as directed by a computer using a downloaded program through wireless interfaces. One of the two sprockets in turn, through additional sprockets/chains/axle devices, is utilized to supply drive torque to a vehicle transmission. The other axle which is connected to the remaining sprocket and through an additional chain/sprocket/axle device operates a direct current generator which produces electricity for charging a battery.

5 Claims, 9 Drawing Sheets

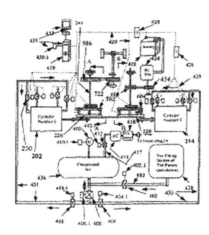

17

Fig. #1

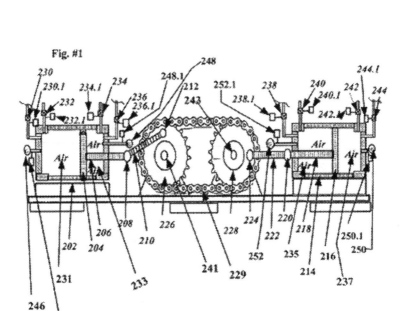

Fig #2

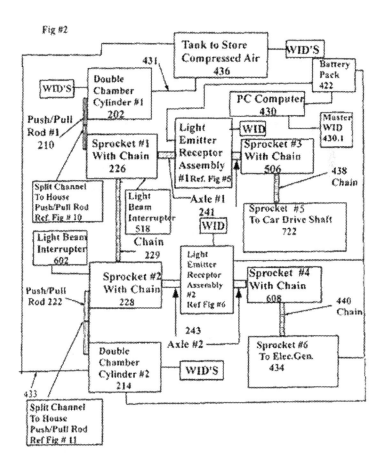

Fig. # 3

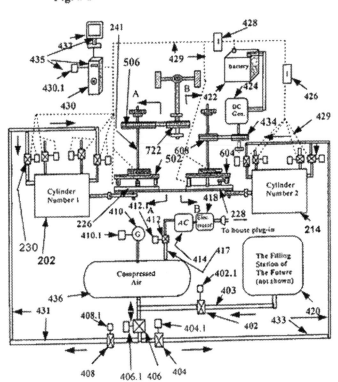

Fig. #4   View A-A

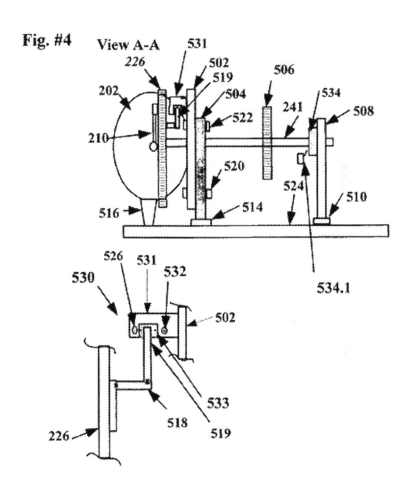

21

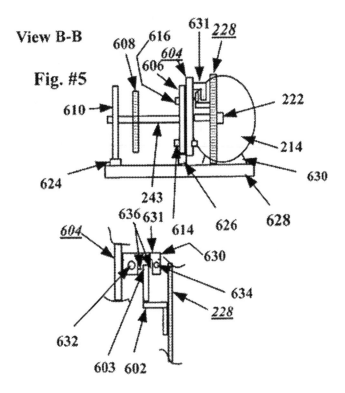

View B-B

Fig. #5

Fig. #6

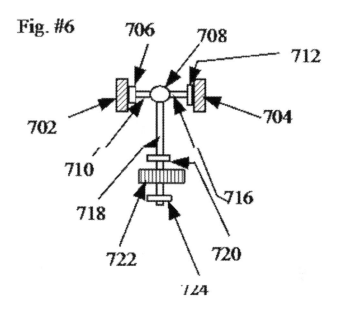

706 708 712
702 704
710
718 716
722 720
724

Fig. #7

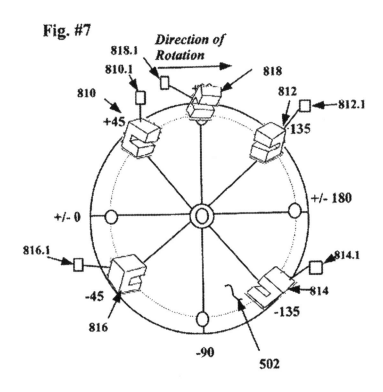

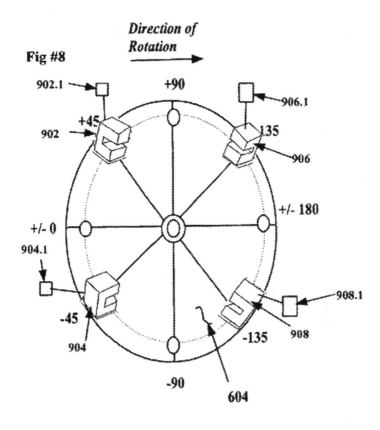

Fig #8

Direction of
Rotation

Fig #9 <u>Step #1</u>

One Complete Operational Cycle

Cylinder 202	Sprocket 226	Sprocket 228	Cylinder 214
I/V 230 Open	Pivot Point 212	Pivot Point 224	I/V 238 Closed
Closed V/V 232	Rotates from/to	Rotates from/to	V/V 240 Open
V/V 234 Open	+45 to +135	+135 to -135	V/V 242
Open			
I/V 236 Closed		Refer to Fig's #1;#2 & #3	I/V 244 Closed
Push Mode			*Vent Mode*

<u>Step #2</u>

I/V 230 Closed	Rotates from/to	Rotates from/to	I/V 238 Closed
V/V 232 Open	+135 to -135	-135 to -45	V/V 240 open
V/V 234 Open			V/V 242 Closed
I/V 236 Closed			I/V 244 Open
Vent Mode			*Push Mode*

<u>Step #3</u>

I/V 230 Closed			I/V 238 Closed
V/V 232 Open	Rotates from/to	Rotates from/to	V/V 240 Open
V/V 234 Closed	-135 to -45	-45 to+45	V/V 242 Open
I/V 236 Open			I/V 244 Closed
Pull Mode			*Vent Mode*

<u>Step #4</u>

I/V 230 Closed	Rotates from/to	Rotates from/to	I/V 238 Open
V/V 232 Open	-45 to +45	+45 to +135	V/V 240 Closed
V/V 234 Open			V/V 242 Open
I/V 236 Closed			I/V 244 Closed
Vent Mode			*Pull Mode*

<u>Step #5</u>

I/V 230 Open	Pivot Point	Pivot Point	I/V 230 Closed
V/V 232 Closed	Rotates from/to	Rotates from/to	V/V 240 Open
V/V 234 Open	+45 to +135	+135 to -135	V/V 242 Open
I/V 236 Closed			I/V 244 Closed
Push Mode			*Vent Mode*

This completes one complete Operational cycle for the invention

1

COMPRESSED AIR PROPULSION SYSTEM

CROSS-REFERENCE TO RELATED APPLICATIONS

This application is a Continuation-In-Part of U.S. Ser. No. 14/256,754 filed Apr. 18, 2014 for THE CAR OF THE FUTURE POWERED BY COMPRESSED AIR.

BACKGROUND OF THE INVENTION

1. Field of the Invention

The present invention relates to a propulsion system, and in particular to a propulsion system using compressed air in which supplies drive torque to a vehicle transmission as well as drive torque to an electronic generator for charging a battery.

2. Description of the Related Art

The proposed invention can be related to providing propulsion of a vehicle in a more efficient and less expensive manner, meanwhile reducing the amount of pollution being released into the atmosphere. The current method is not only wasteful and inefficient but is prone to producing vast amount of pollutants into the atmosphere on a daily basis while not doing useful work at the same time. This vast amount of pollution is believed to be contributing to the global warming of the earth and all of the ills that are attendant with it, like flooding of lower coastal regions etc. The basic reason, as concerns motor vehicles, is the ever increasing congestion on all of the nations roadways whether city streets, state highways or federal freeways. This massive congestion results in prolonged delay of vehicles of all types in trying to reach their destinations. The impressive gas mileages that all of the automobile manufacturing companies extol are meaningless when all of the vehicles during the rush hour spend long periods of time stalled and idling in traffic. Although the electric car is subject to the same traffic delays as all of the other vehicles propelled by a different method such as gas or diesel engine, still the electric car is in the off mode when not in motion. Yet the problem with the electric car and hybrid car, even though to a lesser degree, is the limited amount of storage of electricity available with the current technology in battery manufacturing.

BRIEF SUMMARY OF THE INVENTION

The compressed air propulsion system of the present invention includes a compressed air tank. First and second identical opposing cylinders have a piston connected to a piston rod. Each piston divides the cylinder into two chambers and each chamber is connected to inlet and outlet valves with the inlet valves connected to the compressed air tank. First and second sprockets joined by a common chain are mounted on corresponding first and second axles. The piston rods of the first and second cylinders are connected to corresponding first and second sprockets at a pivot point. Each axle has a fixed disc mounted adjacent to the sprocket. The fixed disc includes plural light beam emitter/receptor devices that send signals to an electronic control unit for controlling the opening and closing of the inlet and outlet valves based on the position of interrupters mounted to each sprocket. The first axle connected to the first sprocket supplies drive torque to a vehicle transmission and the second axle connected to the second sprocket supplies drive torque to an electronic generator which charges a battery.

BRIEF DESCRIPTION OF THE DRAWINGS

FIG. 1 is an end view of the present invention.

FIG. 2 is a block diagram of the present invention.

2

FIG. 3 is a top view of the present invention.

FIG. 4 is a view taken along line A-A of FIG. 3.

FIG. 5 is a view taken along line B-B of FIG. 3.

FIG. 6 is a top view of the drive shaft and its housing.

FIG. 7 shows five brackets affixed to fixed disc 502.

FIG. 8 shows 4 brackets affixed to fixed disc 604.

FIG. 9 is a listing that depicts all of the various modes of the two cylinders for one cycle of the sprockets.

DETAILED DESCRIPTION OF THE INVENTION

Many of the details of the present invention are shown in FIGS. 1 through 5 including a compressed air tank 436 (FIGS. 2 and 3). First 202 and second 214 identical opposing cylinders include a piston connected to a piston rod. First cylinder 202 has piston 204 connected to piston rod 206 while second cylinder 214 has piston 216 connected to piston rod 218.

Each piston is divided into two chambers. As shown in FIG. 1 first cylinder 202 has chambers 231 and 233 while second cylinder 214 has chambers 235 and 237.

Each chamber has inlet valves and outlet or vent valves. Left chamber 231 of first cylinder 202 is provided with inlet valve 230 and vent or outlet valve 232 and pressure gauge 246. Right chamber 233 of first cylinder 202 is provided with inlet valve 236 and vent or outlet valve 234 and pressure gauge 248. Likewise left chamber 235 of second cylinder 214 is provided with inlet valve 238 and vent or outlet valve 240 and pressure gauge 252. Right chamber 237 of second cylinder 214 is provided with inlet valve 244 and vent or outlet valve 242 and pressure gauge 250.

The cylinder chambers are connected to the compressed air tank 436 via inlet valves. As shown in FIG. 3 inlet line 431 connects tank 436 with first cylinder 202 via inlet valves 230 and 236 while inlet line 433 connects tank 436 with second cylinder 214 via inlet valves 238 and 244.

First sprocket 226 and second sprocket 228 are joined by common chain 229. First sprocket 226 is mounted on first axle 241 and second sprocket 228 is mounted on second axle 243.

The piston rods of the first and second cylinders are connected to a corresponding one of the first and second sprockets at a pivot point. As shown in FIG. 1, piston rod 206 of first cylinder 202 is connected to first sprocket 226 at pivot point 212 via joint 208 and push rod 210 while piston rod 218 is connected to second sprocket 228 at pivot point 224 via joint 220 and push rod 222. Since pivot 212 is at 90° and pivot point 224 is at +1-180° the pivot points are 90° out of phase.

Each axle has a fixed disc mounted adjacent to the sprocket. As shown in FIG. 4, first axle 241 has fixed disc 502 mounted adjacent to first sprocket 226. In FIG. 5 second axle 243 has a fixed disc 604 mounted adjacent to second sprocket 228.

Each fixed disc 502 and 604 includes plural light beam emitter/receptor devices 530 (FIG. 4) and 630 (FIG. 5) that send signals to an electronic control unit 435 for controlling the opening and closing of the inlet valves 230 236, 238, 244 and outlet valves 232, 234, 240, 242 based on the position of interrupts 518 and 602 mounted to each sprocket 226 and 228.

The first axle 241 is connected to the first sprocket 226 supplying drive torque to a vehicle transmission (FIG. 3); and the second axle 243 is connected to the second sprocket 228 supplying drive torque to an electronic generator (FIG. 3) which charges a battery.

As shown in FIG. 3 the electronic control unit 435 includes a computer 438, a master wireless interface device 430.1 and a monitor 432. Various wireless interface devices (WID) serve as the wireless link with the master wireless interface device 430.1 which enables the computer 430 to command the opening and closing of inlet and/or outlet valves in a preprogrammed order.

FIG. 1 shows first cylinder 202 having WID 230.1 associated with inlet valve 230, WID 232.1 associated with outlet valve 232, WID 234.1 associated with outlet valve 234, WID 236.1 associated with inlet valve 236, WID 246.1 associated with pressure gauge 246, and WID 248.1 associated with pressure gauge 248. Likewise, second cylinder 214 had WID 238.1 associated with inlet valve 238, WID 240.1 associated with outlet valve 240, WID 242.1 associated with outlet valve 242, WID 244.1 associated with outlet valve 244, WID 250.1 associated with pressure gauge 250, and WID 252.1 associated with pressure gauge 252.

The master wireless interface device 430.1 is the link between the computer 430 and the various switches, pressure gauges and emitter/receptor devices.

The opening and closing of the various switches are dependent on which light beam has been intercepted or blocked by the light beam interrupter. For cylinder 202 that would be 518 from FIG. 4 and for cylinder 214 that would be light beam interrupter 602 of FIG. 5.

Each light beam that is blocked informs the computer to command a particular inlet valve to change to the open or closed mode or for a particular vent valve to change to the open or closed mode.

The changes by valves of both cylinders are made at the same time. The only difference is that each sprocket has its own light beam interrupter attached to the back of it. And since the two cylinders work together it is required that the particular valves open and close in conformance with instructions that have been programmed in the computer program that has been down loaded on the computer hard drive.

The device is intended to work when each piston produces power in sequence, not simultaneously. When one piston is in the push or pull power mode the other piston is in the vent mode.

As stated, the device operates in several modes. First is the start up mode in the clockwise direction, then operation mode. This can be of the push or pull mode by either of the pistons. There is also the shut down mode where the device is brought to a halt. These modes apply only to the two cylinders.

In FIG. 2 compressed air from tank 436 goes through inlet line 431 to first cylinder 202 to drive push rod 210 which rotates first sprocket 226 and downstream sprocket 506 on first axle 241. Then downstream sprocket 506 by means of chain 438 rotates connecting sprocket 722 which connects to a vehicle drive shaft. In the same manner compressed air from tank 436 goes through inlet line 433 to second cylinder 214 to drive push rod 222 which rotates second sprocket 228 and downstream socket 608 on second axle 243. Then downstream socket 608 by means of chain 440 rotates connecting sprocket 434 which connects to electric generator 424. The electric generator supplies power to battery pack 422.

FIG. 3 has another mode that involves the valves and pressure gauges that are involved in introducing air under high pressure into the various tanks from one or more external sources. One would be from a filling station of the future or from a compressor located in the owners garage.

Electric motor 418 operates air compressor 414 which supplies compressed air via line 417 to tank 436 through valve 412. Also, station 420 supplies compressed air via line 403 to tank 436 through valve 402 when valve 406 is closed. Pressure within tank 436 should be around 600 psi while the pressure within cylinders 202 and 214 as regulated by the various pressure gauges is between 10-15 psi.

FIG. 3 also shows connecting sprocket 434 rotating generator axle to supply electricity to DC generator 424 which goes into battery 422 via line 423. DC from the battery 422 goes through inverters 426 and 428 to become AC which then goes through electric lines 429 to supply power to all valves, WIDS, pressure gauges and other equipment that require electricity.

The lower portion of FIG. 4 is a larger view of the emitter/receptor assembly, generally indicated 530, which has a light beam interrupter 518 attached to the back side of first sprocket 226 and bracket 531 attached to fixed disc 502. Bracket 531 houses light beam emitter 526 and light beam receptor 532 having light beam 533 therebetween. As sprocket 226 is rotated finger 519 of light beam interrupter 518 breaks the light beam 533 between emitter 526 and receptor 532, the consequences of which will be described later.

Supports 504 and 508 on base 524 house first axle 241 which is connected to first sprocket 226 and downstream sprocket 506. Flange 514 attaches to support 504 and flange 510 attaches to support 508. Fixed disc 502 is connected to support 504 by bolts 520 and 522. Brake 534 is on axle 241 and is activated by WID 534.1.

The lower portion of FIG. 5 is a larger view of the emitter/receptor assembly, generally indicated 630, which has a light beam interrupter 602 attached to the back side of second sprocket 228 and a bracket 631 attached to fixed disc 604. Bracket 631 houses light beam emitter 632 and light beam receptor 634 having light beam 636 therebetween. As sprocket 228 is rotated finger 603 of light beam interrupter 602 breaks the light beam 636 between emitter 632 and receptor 634, the consequences of which will be described later.

Supports 606 and 610 on base 628 house second axle 243 which is connected to second sprocket 228 and downstream sprocket 608. Flange 626 attaches to support 606 and flange 624 attaches to support 610. Fixed disc 604 is connected to support 606 by bolts 614 and 616.

FIG. 6 is an enlarged view of a portion of FIG. 3 wherein supports 720 and 724 hold housing/drive shaft 718 which engages differential 708. Rotation of sprocket 722 rotates drive shaft 718 which through differential 708 rotates left axle 710 and right axle 716 thereby rotating wheels 702 and 704. Brake 706 is adjacent wheel 702 while brake 712 is adjacent wheel 704.

FIG. 7 shows brackets 810, 812, 814 and 816 attached to fixed disc 502 at angles of +45, +135, −135, and −45 respectfully. Bracket 818 at +90 is de-energized in operational mode and is activated during the shut down mode. As described above for FIG. 4 each bracket houses a light beam emitter and a light beam receptor having a light beam therebetween. WIDs 810.1, 812.1, 814.1, 816.1, and 818.1 are associated with their respective brackets.

FIG. 8 shows the four brackets on fixed disc 604. Bracket 902 is attached at +45°, bracket 904 at 45°, bracket 906 at +135°, and bracket 908 at −135°. WIDs 902.1, 904.1, 906.1 and 908.1 are associated with their respective brackets. As described above for FIG. 5 each bracket houses a light beam emitter and a light beam receptor having a light beam therebetween.

FIG. 9 lists the open and closed positions of the inlet and vent valves as the sprockets 226 and 228 complete one operational cycle. This will be explained in detail later on.

The invention has two power units consisting of two cylinders that are in opposition. Each cylinder has an enclosed piston which divides the cylinder into two chambers. Each chamber of each cylinder is furnished with an inlet valve and an outlet or vent valve.

5

Each inlet valve allows air under pressure from the internal storage tank(s) to flow into the particular chamber in conformance with instructions previously included in a computer program downloaded on the computer hard drive. This pressured air in turn applies force to a piston. Each piston is connected to a push/pull bar that moves right or left with the piston. Each horizontal push/pull bar in turn is connected to a second push/pull bar that pivots at the connection point with its respective horizontal push/pull bar.

Functioning of the individual vent valves is identical to that of any inlet valves. Commands to change from the open or closed mode are transmitted wirelessly through commands from the electronic control unit through electronic pulses.

Any one of the four individual pressure gauges serves only to provide a constant readout to the computer wirelessly of the pressure existing in a particular chamber at a particular moment in time. Pressure gauges play no active role in generating force by the power unit.

The two sprockets rotate in unison due to a common chain that links them together. Each of the two sprockets is identical in size and shape and functions the same way.

Operation of the inlet valves or vent valves in either the open or closed position is determined by instructions from and through the master wireless interface device 430.1 to the individual wireless interface devices that are part of each of the inlet or vent valve assemblies.

The compressed air propulsion system operates in one of three modes. These modes are startup, operation, and shutdown.

Each of the two cylinders operates in one of four modes. The modes are push, vent, pull and vent. These modes are repetitive. When either cylinder is in the push or pull mode the opposite cylinder is in the open vent mode.

Only one inlet valve of the four inlet valves on the cylinders can be in the open position mode at any time during the operation mode. More than one vent valve in either cylinder may be in the open mode at any one time.

Startup Mode

When the system is placed in the startup mode on computer command inlet valve 230 is ordered to the open mode. Pivot point 212 rotates between locations +90 and +135 relative to fixed disk 502. This causes finger 519 of light beam interrupter 518 attached to the back side of sprocket 226 to break light beam 533 housed within bracket 812 at location +135 of fixed disk 502 resulting in inlet valve 230 changing from the open mode to the closed mode.

Pivot point 224 has been resting at location +180 relative to fixed disk 604 during the shutdown mode. Pivot point 224 rotates from +180 to location -135. This causes finger 603 of light beam interrupter 602 attached to the back side of sprocket 228 to break the light beam 636 housed within bracket 908 of fixed disk 604 causing inlet valve 244 to change from the off mode to the on mode. Meanwhile vent valve 242 changes from the on mode to the off.

The power unit of the invention continues to operate on commands transmitted by wireless pulses to the various valves and switches until the next shutdown operation is reached.

Operation Mode

For cylinder 202 the push mode occurs when inlet valve 230 is in the open mode, vent valve 232 is in the closed mode and inlet valve 236 is in the closed mode while vent valve 234 is in the open mode. At the same time inlet valve 244 of

6

cylinder 214 is in the closed mode and vent valve 242 in the right chamber of cylinder 214 is in the open mode. Inlet valve 238 is in the closed mode while vent 240 is in the open mode. Air flows into the left chamber of cylinder 202 and out through vent valve 234 of the right chamber. During the same interval in time air flows in or out of either vent valve 240 or 242 as the two sprockets rotate.

Cylinder 202 is in either of the two vent modes when both inlet valves 230 and 236 are in the closed mode and the two vent valves 232 and 234 are in the open mode. During the same interval in time inlet valves 238 and 244 of cylinder 214 are in the closed mode. Conversely vent valves 240 and 242 are in the open mode. Air may flow in or out of either chamber depending on the direction the particular piston is moving.

Cylinder 202 is in the pull mode when pressure enters chamber 233 through inlet valve 236 which is in the open mode and vent valve 234 located in the same chamber is in the closed mode. Inlet valve 230 is in the closed mode and vent valve 232 is in the open mode. For cylinder 214 both inlet valves 238 and 244 are in the closed mode and vent valves 240 and 242 are in the fully open mode.

Cylinder 214 is in the push mode when inlet valve 244 is in the open mode, vent valve 242 is in the closed mode, inlet valve 238 is in the closed mode and vent valve 240 is in the open mode. At the same time both inlet valves 230 and 236 of cylinder 202 are in the closed mode while both vent valves 232 and 234 are in the fully open mode.

Cylinder 214 is in either of the two vent modes when inlet valves 238 and 244 are in the closed mode and vent valves 240 and 242 are in the fully open mode. Cylinder 202 is in the either the push mode or the pull mode.

Cylinder 214 is in the pull mode when inlet valve 238 is in the open mode, vent valve 240 is in the closed mode, inlet valve 244 is in the closed mode and vent 242 is in the vent mode.

As shown in FIG. 9, cylinder 202 is also in the push mode when pivot point 212 rotates between +45 and +135 relative to fixed disk 502 (FIG. 3 & FIG. 7) and in the vent mode when pivot point 212 rotates between +135 and -135. It is in the pull mode when pivot point 212 rotates between -135 and -45. It is in a second vent mode when pivot point 212 rotates between -45 and +45. Any following rotation repeats itself. The rotation of either sprocket depends on direction of rotation of either push/pull bar depending on which is in the power mode.

Cylinder 214 is in the push mode when pivot point 224 rotates between -135 and -45 of fixed disk 502. It is in one of two vent modes when pivot point 224 rotates between -45 and +45. It is in the pull mode when pivot point 224 rotates between +45 and +135. It is in a second vent mode when pivot point 224 rotates between +135 and -135.

Opening or closing of any of the four inlet valves of the two cylinders or the vent valves thereof depends on pulses transmitted through wireless interfaces devices (WIDs). The pulses generated by the master wireless interface device 430.1 are determined by instructions embedded in the computer program. These pulses are transmitted in a predetermined order to achieve a smooth rotation of the two sprockets. Rotation of the sprockets in turn causes the axles they are mounted on to rotate. One axle 241 serves to provide motion for a vehicle while rotation of a second axle provides rotation of the armature of the DC generator to generate electricity to power the invention.

In operation starting with cylinder 202 in the push mode with inlet valve 230 in the open mode, vent valve 232 in the closed mode, vent valve 234 in the open mode while inlet valve 236 is in the closed mode. In the same interval of time both inlet valve 238 and 244 of cylinder 214 are in the closed

29

mode and vent valves 240 and 242 are in the open mode. Air under pressure is introduced through inlet valve 230 and flows out of right chamber vent valve 234. Air that is in the chambers 235 and 237 flows in or out of either chamber depending on the motion of piston 216.

Each sprocket has a light beam interrupter device (LBI) mounted on the reverse side of the sprocket. Sprocket 226 has LBI 518 (refer to FIG. 4) mounted along a radius extending from the center of axle 241 (refer to FIG. 1) vertically to a position opposite to pivot point 212. LBI 602 (refer to FIG. 5) is mounted along a radius extending from the center of axle 243 (refer to FIG. 1) to a point opposite to pivot point 224).

The finger 519 of LBI 518 breaks in succession the light beams 533 housed in brackets 818, 812, 814 and 816 (refer to FIG. 4 & FIG. 7).

Meanwhile the finger 603 of LBI 602 breaks successively through the light beams 636 housed in brackets 902, 906, 908 and 904 (refer to FIG. 5 & FIG. 8).

For the emitter/receptor assembly 530 mounted on the face of fixed disc 502 (refer to FIG. 3) pulses are generated when the finger 519 of LBI 518 momentarily breaks the light beam existing between the particular emitter/receiver device gap. This also occurs simultaneously when the finger 603 of LBI 602 momentarily breaks the light beam existing between the particular emitter/receptor devices that are mounted on the face of fixed disk 604.

The emitter/receptor housed in bracket 818 is provided with electrical power only during the shutdown mode but is not active in the startup or operational mode.

Blockage of the light beam within bracket 810 at location +45 on fixed disk 502 causes the computer through the master wireless interface device 430.1 to transmit a pulse or pulses that cause inlet valve 230 to change from the closed mode to the open mode. Another pulse generated at the same time causes vent valve 232 to change from the open mode to the closed mode as cylinder 202 changes from the vent mode to the push mode. Other pulses when LBI 602 breaks the light beam at −45 on fixed disk 604 (refer to FIG. 3) causing inlet valve 244 to change from the open mode to the closed mode and another pulse causes vent valve 242 to change from the closed mode to the open mode as cylinder 214 changes from the push mode to the vent mode.

Blockage of the particular light beam of any of the emitter/receptor devices in like manner causes inlet and outlet valves of both cylinders to command the particular inlet or vent valves to open or close as cylinders change from one mode to the next mode.

Shutdown Mode

When the shutdown activating device (not shown) is in the on mode and on the second rotation of pivot point 212 through location +45 of fixed disk 502 inlet valve 230 is changed to the open condition until pressure gauge 246 by pulse verifies that the design pressure of 10 psi in chamber 231 has been reached. At this time inlet valve 230 is placed in the closed mode. This is imposed on inlet valve 230 so that the pressure in chamber 231 decreases as sprocket 226 continues to rotate and chamber 231 increase in volume causing a decrease in pressure. This preprogrammed decrease in pressure results in a decrease in the force exerted on the face of piston 204. When

LBI 518 breaks the light beam E/R/D within 818 the electric current to E/R/D within 818 is placed in the off mode. At the same instant brake 534 mounted on axle 241 is placed in the engaged mode and the rotation of axle 241 is brought to a halt. In the shutdown mode both cylinders rest in the vent mode where inlet valves 230 and 236 of cylinder 202 rest in the closed mode and vent valves 232 and 234 rest in the open mode. Inlet valves 238 and 244 of cylinder 214 rest in the closed mode while vent valves 240 and 242 rest in the open mode.

Although particular embodiments of the present invention have been described and illustrated, such is not intended to limit the invention. Modifications and changes will no doubt become apparent to those skilled in the art, and it is intended that the invention only be limited by the scope of the appended claims.

The invention claimed is:

1. A compressed air propulsion system, comprising:
   a compressed air tank;
   first and second identical opposing cylinders each including a piston connected to a piston rod, each piston dividing the cylinder into two chambers;
   each chamber connected to inlet valves and outlet valves;
   the cylinder chambers connected to the compressed air tank via the inlet valves;
   first and second sprockets joined by a common chain, the sprockets mounted on corresponding first and second axles;
   the piston rods of the first and second cylinders are connected to a corresponding one of the first and second sprockets at a pivot points;
   each axle having a fixed disc mounted adjacent to the sprocket, wherein the fixed disc includes plural light beam emitter/receptor devices that send signals to an electronic control unit for controlling the opening and closing of the inlet valves and outlet vales based on the position of interrupters mounted to each sprocket;
   the first axle connected to the first sprocket supplying drive torque to a vehicle transmission; and
   the second axle connected to the second sprocket supplying drive torque to an electronic generator which charges a battery.

2. The compressed air propulsion system of claim 1 wherein:
   each cylinder has 2 inlet valves, 2 outlet valves, and a pressure gauge.

3. The compressed air propulsion system of claim 1 wherein:
   each emitter/receptor device has a light beam interrupter attached to the back side of a sprocket.

4. The compressed air propulsion system of claim 3 wherein:
   a bracket attached to a fixed disc which houses a light beam emitter, and a light beam receptor having a light beam therebetween.

5. The compressed air propulsion system of claim 4 wherein:
   rotation of the sprocket causes a finger on the light beam interrupter to break the light beam between the light beam emitter and the light beam receptor.

* * * * *

GLOSSARY

Drawings 1 through 9

Compressed Air Propulsion System
Fig. #2
#200 Power Unit
#202 Cylinder #1
#204 Piston #1
#206 Push/Pull Rod
#208 Pivot Point
#210 Push/Pull Rod
#212 Pivot Point
#214 Cylinder #2
#216 Piston #2
#218 Push/Pull Rod
#220 Pivot Point
#222 Push/Pull Rod
#224 Pivot Point
#226 Sprocket #1
#228 Sprocket #2
#230 Inlet Valve Cyl. #1
#230.1 Wireless Interface **Device (W/I/D)**
#232 Vent Valve
#232.1 W.I.D
#234 Vent Valve
 #234.1 W/I/D
#236 Inlet Valve Cyl. ##236.1 W/I/D#238 Inlet Valv. #2
#238.1W/I/D
#240 Vent Valve Cyl. #2
#240.1 W/I/D
#242 Vent Valve Cyl. #2
#242.1 W/I/D
#244 Inlet Valve Cyl. #2

#244.1 W/I/D
#246 Pressure Gage Cyl. #1
#246.1W/I/D
#248 Pressure Gage Cyl.#1
#248.1 W/I/D
#250 Pressure Gage Cyl. #2
#250.1W/I/D
#252 Pressure Gage Cyl. #1
#252.1 W/I/D
#254 Split Ring Piston
#256,1 & .2 Face Plates
#258.1 & .2 Bolts
#260 Pin
#260 Hub
Fig #3
#402 2 Way Valve
#402.1 W/I /D
#404 2 Way Valve
#404.1 W/I/D
#406 2 Way Valve
#406.1 W/I/D
#408 2 Way Valve
#408.1 W/I/D
#410 Pressure Gage
#410.1 W/I/D
#412 Two Way Valve
#412.1 W/?D
#414 Air Compressor
#418 Electric Motor
#422 Battery Pack
#424 Elec. Gen.
#426 Inverter
#428 Inverter

#430 Computer
#430.1 Master Wireless
Interface Device
#432 Monitor
#434 Pulley or Sprocket
#436 Compressed
Air Tank/or Tanks
Fig. #5
#502 Fixed Disk
#504 Support Strut
#506 Pulley or Gear
#508 Support Strut
#510 Flange
#512 Axle
#514 Flange
#516 Strut
#518 Light Beam
Interrupter
#520 Bolt
#522 Bolt
#524 Base
#530 Light Beam

Emmiter
#532 Light Beam
Receptor
#533 Light Beam (typ.)
Fig. 6
#602 Light Beam
Interruptor
#604 Fixed Disk
#606 Support Strut
#608 Pulley or Gear
#610 Strut
#243 Axle
Fig. 7 (Not a part of this
Contract)
(operates another DC Generator)
#702 Wheel
#704 Wheel
#706 Air Brake
#708 Air Brake
#710 Axle
#712 Air Brake

**DRAWINGS NOT INCLUDED
IN THE PATENT**

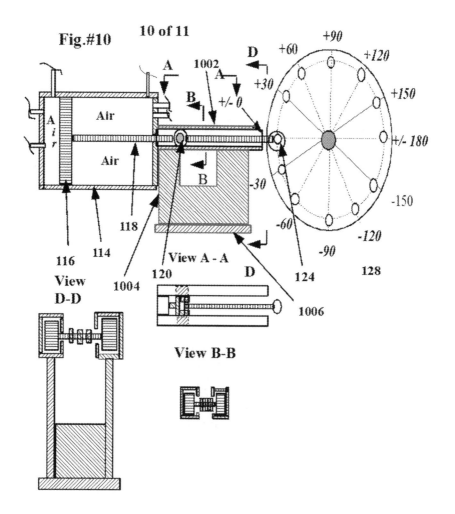

Fig.#10

View A - A

View B-B

View D-D

Glossary Fig # 10
1002 Split channel
1004 Web
1006 Flange
The 1000 structure is intended to react against
the vertical force component of the applied force.

Fig. #11

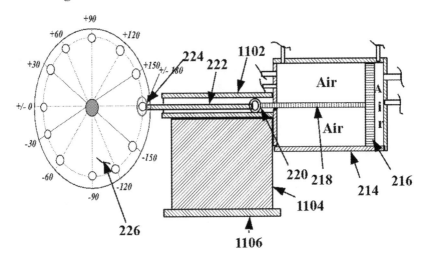

Glossary Fig # 11
1102 Split Channel
1104 Web
1106 Flange
Note: The 1100 structure is intended
to react to the vertical force
component of the
applied force

Fig # 12 The Design Configuration

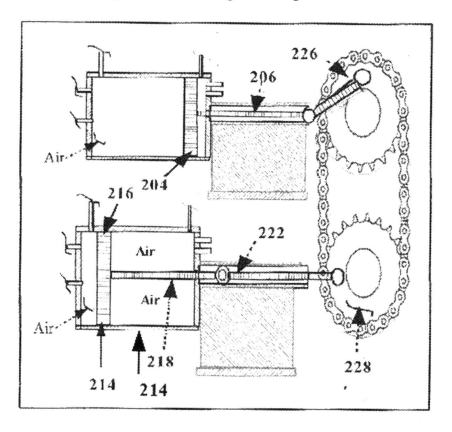

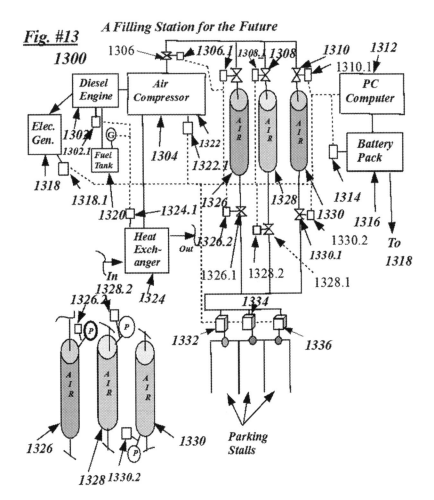

Fig. #13
1300

A Filling Station for the Future

Glossary not provided

AN ALTERNATE CONFIGURATION UTILIZING CAMS TO OPERATE THE INLET AND VENT VALVES

Fig. #4A View A-A

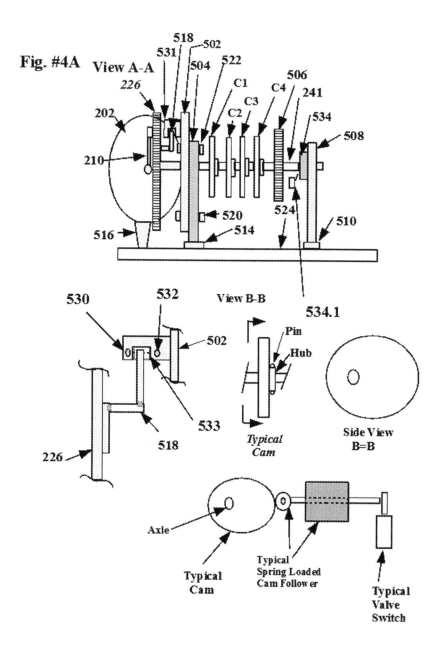

View B-B

Typical Cam

Side View B=B

Axle

Typical Cam

Typical Spring Loaded Cam Follower

Typical Valve Switch

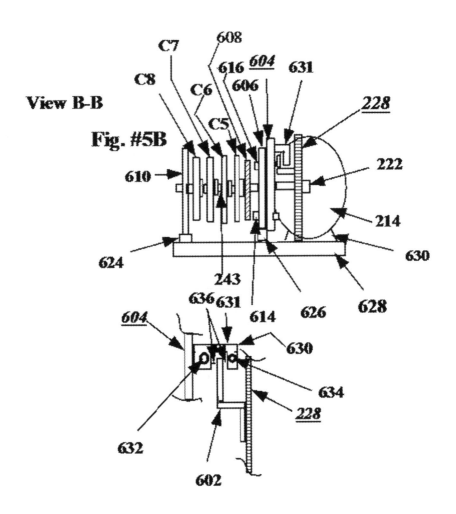

View B-B

Fig. #5B

Fig #9 (Cam Operated)

One Complete Operational Cycle

Step #1

Cylinder 202	Sprocket 226	Sprocket 228	Cylinder 214
I/V 230 (C1/Ext)	Pivot Point 212	Pivot Point 224	I/V 238 (C5/Ret)
V/V 232 (C2/ Ret)	Rotates from/to	Rotates from/to	V/V 240 (C6/Ext)
V/V 234 (C3/Ext)	+45 to +135	+135 to -135	V/V 242 (C7/Ext)
I/V 236 (C4/Ret)	Refer to Fig's #1;#2 & #3		I/V 244 (C8/Ret)
Push Mode			*Vent Mode*

Step #2

I/V 230 (C1/Ret)			I/V 238 (C5/Ret))
V/V 232 (C2/Ext)	Rotates from/to	Rotates from/to	V/V 240 (C6/Ext)
V/V 234 (C3/Ext)	+135 to -135	-135 to -45	V/V 242 (C7/Ret)
I/V 236 (C4/Ret)			I/V 244 (C8/Ext)
Vent Mode			*Push Mode*

Step #3

I/V 230 (C1/Ret)			I/V 238 (C5/Ret0
V/V 232 (C2/Ext))	Rotates from/to	Rotates from/to	V/V 240 (C6/Ext)
V/V 234 (C3/Ret)	-135 to -45	-45 to+45	V/V 242 (C7/Ext)
I/V 236 (C4/Ext)			I/V 244 (C8/Ret)
Pull Mode			*Vent Mode*

Step #4

I/V 230 (C1/Ret)	Rotates from/to	Rotates from/to	I/V 238 (C5/Ext)
V/V 232 (C5/Ext)	-45 to +45	+45 to +135	V/V 240 (C6/Ret)
V/V 234 (C3/Ext)			V/V 242 (C7/Ext)
I/V 236 (C4/Ret)			I/V 244 (C8/Ret)
Vent Mode			*Pull Mode*

Step #5

I/V 230 (C1/Ext)	Pivot Point	Pivot Point	I/V 238 (C5/Ret)
V/V 232 (C3/Ret)	+45 to +135	+135 to -135	V/V 240(C6/Ext)
V/V 234 (C2/Ext)	Rotates from/to	Rotates from/to	V/V 242(C7/Ext)
I/V 236 (C4/Ret)			I/V 244(C8/'Ret)
Push Mode			*Vent Mode*

Note: "Ret" stands for Retracted or Closed concerning valves.

"Ext" stands for Extended or Open concerning valves.

"C" Stands for Cam. Each Cam has a Cam Follower (CF)

This completes one complete Operational cycle for the invention

SPECIFICATIONS FOR THE CAM OPERATED SYSTEM

Note: Not a part of the Patent

The Cam Operated System Specifications

A different method of switching the inlet and vent valves involves using individual cams in contact with spring loaded cam followers. Refer to the drawings 4A and 5B.

By modifying drawings Fig #4 & Fig.5 to Fig.#4A & 5B, by the addition of four (4) cams identified as C1, C2, C3 and C4 and four other cams C5, C6, C7 and C8, an alternate solution is provided.

As shown in the drawing of a typical cam with a follower, which is in contact with a typical valve switch, indicates how the rotation of each Cam can cause a particular valve to open or close as the Cam rotates through discrete intervals in operation.

By mounting the individual Cam's on axle 241 of drawing 4A, or mounting them on axle 243 of drawing 5B results in the cam's rotating with the individual axles which rotate with sprockets 226 & 228.

By aligning each Cam with its longitudinal axis at the proper angle with the horizontal as a reference can effectively result in the individual cam follower placing the individual valve in the open mode at the proper time and duration in the device rotational operational sequence.

It is envisioned that the two sets of four cams, with there attendant spring loaded cam followers, shall be assembled with the appropriate spring loaded inlet and vent valve switches.

The computer is aware of the location of each Light Beam Interrupter (LBI) at any instant in time during operation and is capable of placing Outlet Valve 406 in the Open mode, meantime cycling valve 408 or 404 to introduce and maintain the required design air pressure in the active chamber of the proper cylinder which is active at the moment. Since, with this concept valve opening is achieved by cam action, the Computer action through WID in the current concept is replaced for these eight valves. The valve action is changed by replacing the WID inlet and vent valves with a different type of valve. The replacement valves all are spring loaded or are furnished with a spring loaded switch.

If any of the eight (8) valve switch springs are in the uncompressed mode the particular valve is in the closed mode. If however, the particular valve spring is compressed by the force applied by the valve follower the valve is in the open mode during that particular interval.

.The role of the Computer is limited to placing outlet valve 406 in the open mode concurrently with outlet valve 408 also being in the open mode when inlet valve 230 or 236 is the open mode. Outlet valve 404 remains in the closed mode during this interval of time. The Computer still monitors the pressure in each chamber

When inlet valves 238 or 244 are in the open mode, outlet valve 406 remains in the open mode while outlet valve 404 changes from the closed mode to the open mode while outlet valve 408 changes from the open mode to the closed mode.

Valve opening is to be achieved indirectly by cam action. If a cam follower exerts a force on a particular spring loaded switch it is considered to be in the **Extended Mode.** If it, the follower, is in contact with the particular switch but does not exert a force to override the valve spring it, the follower, is said to be in the **Retracted Mode.**

Cam C1 with its cam follower CF1 interfaces with inlet valve 230 through its spring loaded switch VSW 1(typ. not shown). Cam 2 with its cam follower CF2 interfaces with vent valve 232 through its spring loaded switch VSW 2. Cam C3 with its cam follower CF3 interfaces with vent valve 234 through its spring loaded switch VSW3. Cam C4 with its cam follower CF4 interfaces with inlet valve 236 through its spring loaded switch VSW 4.

Cam 5 with its cam follower CF5 interfaces with inlet valve 238 through its spring loaded switch VSW 5 (not shown). Cam C6 with its cam follower CF6 interfaces with vent valve 240 with its spring loaded switch VSW6 Cam C7 with its cam follower CF7 interfaces with vent valve 242 through its spring loaded switch VSW7. Cam C8 with its cam follower CF8 interfaces with inlet valve 244 through its spring loaded switch VSW8.

Cams C1, C2, C3 and C4 rotate with sprocket 226 and are fixed to axle 241 Cam followers CF1, CF2, CF3 and CF4 do not rotate but are fixed and limited to longitudinal motion and are arrayed in a fixture which brings each one into contact with its particular valve switch without exerting a force on that switch Cam followers C5, C6, C7 and C8 do not rotate but are fixed and limited to longitudinal motion and are arrayed in a fixture which brings each one in contact with its particular valve switch.

Cams C5, C6, C7 and C8 rotate with sprocket 228 and are fixed to axle 243. These four cams are arrayed in a fixture and are each one is in contact with its respective cam follower.

If the individual cam is configured in such a way that the tip of the cam follower is in contact with the valve switch, but does not apply any force to the switch, then the valve spring remains in the uncompressed mode and the valve remains in the closed mode throughout that particular interval of a full rotation.

PATENT US 8215111

Title: Electrical generation from explosives

| United States Patent | 8,215,111 |
| Richey | July 10, 2012 |

Electrical generation from explosives

Abstract

An electrical generator converts the high blast pressures of explosives into useful electricity by capturing the explosive gases and using the high gas pressures to alternately push water hydraulically between two tanks and through water turbines connected to DC electric generators. Water expelled through a water turbine from one tank is used to fill the other tank. Batteries can be used to store the electrical energy generated, and inverters followed by transformers convert the DC electric from the turbine-generators to 110-VAC, 220-VAC, and 440-VAC. A microcomputer controller connected to various sensors and solenoid valves coordinates the timing and routing of the detonation of explosives, tank pressures, venting, valving, and load control.

Inventors: **Richey; Robert John** (Campbell, CA)
Family ID: 46395800
Appl. No.: 12/386,822
Filed: April 23, 2009

Related U.S. Patent Documents

Application Number	Filing Date	Patent Number	Issue Date
61100915	Sep 29, 2008	<TD< TD>	
		<TD< TD>	

Current U.S. Class:	60/512; 60/325; 60/415
Current CPC Class:	F03B 13/00 (20130101)
Current International Class:	F01B 29/00 (20060101); F16D 31/00 (20060101); F16D 39/00 (20060101)
Field of Search:	;60/369,484,516,632,634,675,914

References Cited [Referenced By]

U.S. Patent Documents

178925	June 1876	Hardy
3611723	October 1971	Theis
3648458	March 1972	McAlister
3650572	March 1972	McClure
4301774	November 1981	Williams
5551237	September 1996	Johnson
5647734	July 1997	Milleron
5713202	February 1998	Johnson
5865086	February 1999	Petichakis P.
6182615	February 2001	Kershaw
6739131	May 2004	Kershaw
7531908	May 2009	Fries et al.

Foreign Patent Documents

WO 01/31197	Mar 2001	WO

Primary Examiner: Denion; Thomas
Assistant Examiner: Jetton; Christopher

Parent Case Text

RELATED APPLICATIONS

This Application claims benefit of U.S. Provisional Patent Application, A UNIQUE METHOD OF GENERATING

ELECTRICITY USING EXPLOSIVE SUBSTANCES AS A POWER SOURCE, Ser. No. 61/100,915, filed Sep. 29, 2008.

Claims

The invention claimed is:

1. A generator system, comprising: a high pressure gas tank providing for the capture and confinement of gases generated by an explosive cartridge; a magazine and breach connected to the high pressure gas tank, and providing for the operation of said explosive cartridge; a pair of interconnected liquid tanks connected to receive gases routed from the high pressure gas tank, said liquid tanks containing a liquid and interconnected such that said liquids within flow in a circuit between the liquid tanks; a high pressure turbine connected to be driven by said liquids flowing between the liquid tanks; a low pressure gas tank connected to the pair of interconnected liquid tanks, and providing for the capture and confinement of gases vented from the interconnected liquid tanks; a second pair of interconnected liquid tanks connected to receive gases routed from the low pressure gas tank, said liquid tanks containing a liquid and interconnected such that said liquids within flow in a circuit between the liquid tanks; a low pressure turbine connected to be driven by said liquids flowing between the second pair of interconnected liquid tanks; an electric generator connected to be driven by the high pressure and low pressure turbines and able to produce electrical power; and a controller to operate valves and to coordinate the timing such that gas pressure from the pressurized gas tank is alternately routed to each liquid tank until the liquid inside is pushed out into the other; wherein energy from said explosive cartridge is converted into electrical power.

2. The system of claim 1, further comprising: a high-water float switch and a low-water float switch disposed in each of the

liquid tanks and connected to the controller; wherein the controller is enabled to maintain the liquid levels within each pair of interconnected liquid tanks over an operational range.

3. The system of claim 1, further comprising: a liquid inlet valve providing a controlled input for each of the pair of interconnected liquid tanks that is connected in a circuit to receive liquids from the other liquid tank in the pair.

4. The system of claim 1, further comprising: a liquid outlet valve providing a controlled output for each of the pair of interconnected liquid tanks that is connected in a circuit to transmit liquids to the other liquid tank in the pair.

5. The system of claim 1, further comprising: a liquid inlet valve providing a controlled input for each of the pair of interconnected liquid tanks that is connected in a circuit to receive liquids from the other liquid tank in the pair; a liquid outlet valve providing a controlled output for each of the pair of interconnected liquid tanks that is connected in a circuit to transmit liquids to the other liquid tank in the pair; and a high-water float switch and a low-water float switch disposed in each of the liquid tanks; wherein the liquid inlet valve and liquid outlet valve are controlled by the controller according to signals obtained from the high-water float switch and a low-water float switch.

6. A generator system, comprising: a pressurized-gas tank providing for the capture and confinement of gases generated by an explosive cartridge; a magazine and breach connected to the pressurized-gas tank, and providing for the operation of said explosive cartridge; a pair of interconnected first and second liquid tanks connected to receive gases routed from the pressurized-gas tank, said liquid tanks containing a liquid and interconnected such that said liquids within flow in a circuit between the liquid tanks; a first and a second liquid inlet valve providing a controlled input for each of the pair of interconnected liquid tanks that is connected in a circuit to receive liquids from the other liquid tank in the pair; a first and a second liquid outlet valve providing a controlled output for each of the pair of interconnected liquid tanks that is connected in a circuit to transmit liquids to the other liquid

tank in the pair; a high-water float switch and a low-water float switch disposed in each of the first and second liquid tanks; a first turbine connected to be driven by said liquids flowing from said first liquid tank to said second liquid tank; a second turbine connected to be driven by said liquids flowing from said second liquid tank to said first liquid tank; an electric generator connected to be driven by at least one of the first and second turbines and able to produce electrical power; and a controller to operate the magazine and breach, and first and a second liquid inlet valve, and first and a second liquid outlet valves to coordinate their timing, such that gas pressure from the pressurized-gas tank is alternately routed to each liquid tank until the liquid inside is pushed out into the other, and according to signals obtained from the high-water float switch and low-water float switch; wherein energy from said explosive cartridge is converted into electrical power.

7. A method of converting explosives energy into electrical power, comprising a computer program in software or firmware executed by a conventional microcomputer, with data inputs from sensors and switches digitized for processing, user inputs used to make process control decisions, and outputs to electro-mechanical solenoids to operate gas and hydraulic valves, comprising: closing pressure tank and water tank vents, and closing water tank inlet valves; raising a gas pressure in a pressure tank up to an operating level by firing explosive cartridges as needed; checking a water level inside a water tank-A and if it's at its maximum operating level, beginning a hydraulic cycle, wherein a gas inlet valve-A is opened, a gas vent valve-A is closed, and a water outlet valve-A to an associated turbine-A is opened; wherein, a gas pressure let in can push water out through said outlet valve-A until the water level reaches a minimum, and said outlet valve-A is closed, said gas inlet valve-A is closed, and gas vent valve-A is opened, and water inlet valve-A is opened to receive water from water tank-B; and checking a water level inside a water tank-B and if it's at its maximum operating level, beginning a hydraulic cycle, wherein a gas inlet valve-B is opened, a gas vent valve-B is closed, and a water outlet

valve-B to an associated turbine-B is opened; wherein, a gas pressure let in can push water out through said outlet valve-B until the water level reaches a minimum, and said outlet valve-B is closed, said gas inlet valve-B is closed, and gas vent valve-B is opened, and water inlet valve-B is opened to receive water from water tank-A.

8. The method of claim 7, further comprising: if a user is not requesting a stop of operations, the process repeats in a loop.

9. The method of claim 7, further comprising: if a user is requesting a stop of operations, closing gas inlet pressure valves to water tank-A and tank-B, opening gas vents, and closing said water outlet valves to the turbines; wherein, residual gas pressures inside said pressurized tank is let down if another use cycle is not scheduled immediately.

Description

BACKGROUND OF THE INVENTION

1. Field of the Invention

The present invention relates to electrical power generation, and in particular to devices and methods for converting the gas pressure generated by explosives into electricity.

2. Description of the Prior Art

Useful electrical energy does not exist in nature and it must be converted from other available energy forms such as gasoline, diesel, coal, natural gas, geothermal, steam, hydro, solar, etc. Some of these energy sources are expensive, some are highly pollutant, some are difficult to convert efficiently, and some are not very portable as is needed in vehicles.

Explosives like gunpowder; Cordite, Ballistite, and Poudre-B smokeless powders; trinitrotoluene (TNT), Dynamite; nitroglycerin; Tovex and other water gel explosives; etc., release a lot of energy in a very rapid pulse. Explosives usually have less potential energy than petroleum fuels, but their high rate of energy release produces large blast pressures. TNT has a detonation velocity of 6,940 m/s compared to 1,680 m/s for the detonation of a pentane-air mixture, and the 0.34-m/s stoichiometric flame speed of gasoline combustion in air. Explosives are classified as "low" or "high" explosives according to their rates of decomposition. Gunpowder is a low explosive, while TNT is a high explosive. Low explosives burn rapidly or deflagrate, while high explosives detonate.

The energy released includes high levels of heat, light, and gas pressure. These are all quickly dissipated if not captured or otherwise contained. For example, at 15.degree. C. the volume of gas produced by the explosive decomposition of one mole of nitroglycerin, becomes, V=(23.64 liter/mol)(7.25 mol)=171.4 liters. The molar volume of an ideal gas at 15.degree. C. is about 23.64 liters. The potential of an explosive is the total work that can be performed by the gas generated by the explosion. If uncontained, it expands adiabatically from its original volume until its pressure is reduced to atmospheric pressure and its temperature to ambient.

In the nitroglycerin reaction, C.sub.3H.sub.5(NO.sub.3)3. fwdarw.3CO.sub.2+2.5H.sub.2O+1.5N.sub.2+0.25O.s- ub.2, the products are carbon dioxide, water, nitrogen, oxygen, and heat. Therefore, a relatively small solid or liquid volume is converted into a very large volume of relatively benign gases. Nitroglycerin explosions are relatively clean, compared to TNT which is poisonous and produces a lot of carbon soot in its reaction.

Firearms and artillery use the gas pressure generated by the detonation of smokeless powder to accelerate bullets and projectiles to very high muzzle velocities on the order of 2,000+ feet per second. Sticks of explosives are detonated in holes drilled into geologic deposits to fracture the ores and make removing the material as easy as scooping up the pieces.

What is needed is a device and method to convert explosive energy into a more useful form of electrical energy as used in homes and industry.

SUMMARY OF THE INVENTION

Briefly, an electrical generator embodiment of the present invention converts the high blast pressures of explosives into useful electricity by capturing the explosive gases and using the high gas pressures to alternately push water hydraulically between two tanks and through water turbines connected to DC electric generators. Water expelled through a water turbine from one tank is used to fill the other tank. Batteries can be used to store the electrical energy generated, and inverters followed by transformers convert the DC electric from the turbine-generators to 110-VAC, 220-VAC, and 440-VAC. A microcomputer controller connected to various sensors and solenoid valves coordinates the detonation of the explosives, tank pressures, venting, valving, and load control.

These and other objects and advantages of the present invention will no doubt become obvious to those of ordinary skill in the art after having read the following detailed description of the preferred embodiments which are illustrated in the various drawing figures.

IN THE DRAWINGS

FIG. 1 is a simplified functional block diagram of a single-stage electrical generator embodiment of the present invention that cycles pressurized water between two tanks and through two sets of water turbines;

FIG. 2 is a flowchart diagram of an electrical generator method embodiment of the present invention to cycle pressurized water between two tanks and two water turbines, as in FIG. 1;

FIG. 3 is a functional block diagram of a single-stage electrical generator embodiment of the present invention that cycles pressurized water between two tanks and through two water turbines like that of FIG. 1, but that reduces duplication of the DC generators and inverters, and some of the valving; and

FIG. 4 is a simplified functional block diagram of a two-stage electrical generator embodiment of the present invention that uses explosive gases to pressurize water, and then uses pressurized hydraulics to spin electrical generators with hydraulic motors and turbines.

DETAILED DESCRIPTION OF THE PREFERRED EMBODIMENT

In FIGS. 1-4 and the following text, some of the more conventional and routine elements commonly used with gas and hydraulic valves, pressure tanks, plumbing, and process control systems are not shown or described. For example, inspection ports and drains for water tanks, safety relief valves, check valves, nozzles for turbines, gearboxes and pulleys, wireless interfaces, wiring, etc. The components like these that should be used are engineering choices and are routinely stocked and installed by technicians. The critical and unusual combinations and their interrelationships are described here in detail.

FIG. 1 represents a single-stage electrical generator system embodiment of the present invention, and is referred to herein by the general reference numeral 100. Generator system 100 produces electrical power suitable for homes, businesses, industry, and the utility grid from the explosive energy captured from cartridges 102 loaded in a magazine 104 and fired in a breach 106. Cartridges 102 should include low explosives that burn clean and soot-free, and the chemical reactions should not produce any dangerous gases or byproducts. For example, nitroglycerin reactions only produce carbon dioxide, water, nitrogen, oxygen, and heat. The heat actually helps increase the gas pressures up to operating levels and should not be wasted or exhausted until the maximum in work has been extracted.

The heated gaseous explosive products are passed through a check valve 108 to a pressurized-gas tank 110. A pressure safety valve (PSV) 112 provides relief if the internal pressures exceed a safe maximum. A pressure sensor (P) 114 measures the tank pressures for a microcomputer controller 120. In some installations the pressure readings will be reported wirelessly, in others a simple 4-20 milliamp process control loop can be used.

Microcomputer controller 120 coordinates all the timing and valve control needed to operate generator system 100 and keep it safe. It uses readings from pressure sensor (P) 114 to determine when more cartridges 102 need to be loaded in magazine 104 and fired in breach 106, and it controls the actual firing. Microcomputer controller 120 also decides when and which gas pressure inlet valve 122 and 124 should be opened and closed for pressurized water tank-A 126 and pressurized water tank-B 128.

Pressurized water tank-A 126 and pressurized water tank-B 128 are not simultaneously pressurized, the pressure applied to them is alternated by gas pressure inlet valves 122 and 124 under control of microcomputer controller 120. What's important to the timing is the water levels inside the tanks, there are minimum and maximum operating levels that must be respected. Water inside one tank needs to flow out into the other tank through a water turbine, and the water cannot flow if the receiving water tank is pressurized at the same time.

In FIG. 1, an outlet valve-A 130 is opened to pass pressured liquid water (L) to a water turbine-A 134. Similarly, an outlet valve-B 132 is opened to pass pressured liquid water (L) to a water turbine-B 136. The liquid water returns from water turbine-B 136 through an inlet valve-A 138 back to water tank-A 126. Liquid water from water turbine-A 134 passes on through to inlet valve-B 140 to water tank-B 128.

The minimum and maximum operating levels of water that circulate between water tank-A 126 and water tank-B 128 are set by float switches (L, H) 142 and 144 for water tank-A 126, and by float switches (L, H) 146 and 148 for water tank-B 128. These float switches are connected to microcomputer controller 120, and the readings are used to determine when to open and close outlet valve-A 130, outlet valve-B 132, inlet valve-A 138, and inlet valve-B 140. The float switch connection could be done wirelessly, and a local loop could be included to automatically close, for example, water outlet valve-A 130 when minimum level float switch 142 senses low water.

Each water tank-A 126 and water tank-B 128 should be equipped with a water to add make up water, and to drain water completely, e.g., during maintenance.

In FIG. 1, water turbine-A 134 is mechanically connected by a rotating shaft to drive a DC electrical generator 150. The DC electrical power produced could be stored in batteries, and it is converted to AC electrical power by an inverter 152. Similarly, water turbine-B 136 is mechanically connected by a rotating shaft to drive another DC electrical generator 154. The DC electrical power produced could be stored in the same batteries, and it can also be independently converted to AC electrical power by an inverter 156. The voltage outputs of inverters 152 and 156 can be stepped-up or stepped-down by conventional transformers as needed, e.g., to 110-VAC, 220-VAC, and 440-VAC.

Microcomputer controller 120 is connected to sense the electrical loads placed on inverters 152 and 156, and uses the information to control how much pressurized water is needed to be passed through water turbine-A 134 and water turbine-B 136 to keep the overall operation in balance.

Once the pressurized gas inside the water tanks has done its job pushing out the water down to its minimum operating level, the residual pressurized gas can be vented out. For water tank-A 126, a vent valve-A 160 is used, and for water tank-B 128, a vent valve-A 162 is used. The residual gas pressures can be high enough to do useful work in a second stage generator. But any back pressure caused by the use of later stages can reduce the efficiency of the earlier stages by reducing the differential pressures between the pressurized tank and the vented one.

In operation, falling water levels inside the water tanks can be used by the minimum-level float switches 142 and 146 to trigger closed the associated water outlet valves 130 and 132. This, in turn can be used to trigger closed the gas pressure inlet valves 122 and 124, and to trigger open the gas pressure vent valves 160 and 162. Similarly, the maximum-level float switches 144 and 148 can be used to trigger closed the water inlet valves 138 and 140.

Pressurized water tank-A 126 and pressurized water tank-B 128 would normally be equipped with various conventional items not shown in FIG. 1. For example, inspection ports, drain valves, pressure gauges, pressure safety valves to release excess pressure, and a water make-up input to replace lost water.

Microcomputer controller 120 can increase and decrease the torque outputs of water turbine-A 134 and water turbine-B 136 by sending modulation controls to nozzle controls 170 and 172. Alternatively, water outlet valve-A 130 and water outlet valve-B 132 could be continuously adjustable, instead of simple fully open, fully closed solenoid types. Such torque modulation would be necessary in some applications to balance the power being generated with the loads applied. In such case, inverters 152 and 156 would also be required to provide load measurements to microcomputer controller 120.

FIG. 2 represents an electrical generator method embodiment of the present invention to cycle pressurized water between two tanks and two water turbines, as in FIG. 1. Such method is referred to herein by the general reference numeral 200. Method 200 is implemented as a computer program in software or firmware executed by a conventional microcomputer, e.g., microcomputer controller 120 (FIG. 1). Data inputs from sensors and switches are digitized for processing, user inputs are used to make process control decisions, and outputs to electro-mechanical solenoids are used to operate gas and hydraulic valves.

Method 200 includes three phases of operation: (1) startup, (2) power generation, and (3) shutdown. During startup, the operational pressures and valve settings must be initialized. During power generation, the gas pressure generated by the explosive cartridges must be switched between the two water tanks according to the respective water levels inside each. The amount of water forced between the water tanks and through the water turbines must be balanced with the electrical loads being placed on the system. During shutdown, the cycling must be stopped and the pressures relieved by opening the various vents.

Specifically, method 200 includes a step 202 for checking to see if the user wants to begin operation. If so, a step 204 closes the pressure tank and water tanks vents, and closes the inlet valves to the water tanks. A step 206 gets the gas pressure in the pressure tank up to operating levels by firing explosive cartridges as needed. A step 208 checks the water level inside water tank-A and if it's at its maximum operating level, a hydraulic cycle can begin. The gas inlet valve-A is opened, the gas vent valve-A is closed, and the water outlet valve-A to the associated turbine-A is opened. The gas pressure let in will push the water out through the outlet valve-A. When the water level reaches minimum, the outlet valve-A is closed. The gas inlet valve-A is closed, and the gas vent valve-A is opened. The water inlet valve-A is opened to receive water from water tank-B.

A step 210 checks the water level inside water tank-B and if it's at its maximum operating level, a hydraulic cycle can begin. The gas inlet valve-B is opened, the gas vent valve-B is closed, and the water outlet valve-B to the associated turbine-B is opened. The gas pressure let in will push the water out through the outlet valve-B. When the water level in water tank-B reaches minimum, the outlet valve-B is closed. The gas inlet valve-B is closed, and the gas vent valve-B is opened. The water inlet valve-B is opened to receive water from water tank-A.

If the user is requesting a stop of operations, a step 212 passes control to a step 214. Otherwise, the process repeats in a loop back to step 206. Step 214 closes the gas inlet pressure valves to water tank-A and tank-B, opens the vents, and closes the water outlet valves to the turbines. Residual gas pressures inside the pressurized tank may be let down if another use cycle is not expected immediately.

FIG. 3 illustrates a single-stage system 300 that eliminates some of the duplication of the major components appearing in FIG. 1. System 300 assumes that when the water level in a water tank is below minimum, e.g., as detected by a low-water float switch, the water outlet valve should be closed. Similarly, when the water level in a water tank is above maximum, e.g., as detected by a high-water float switch, the water inlet valve should be closed. The gas inlet valve to a water tank can only be open if the gas vent is closed. The gas inlet valve to the water tank must be closed if the gas vent is open. The mechanisms implemented to enforce such logic can be built with relay logic, software, IC logic gates, and mechanical interlocks.

System 300 is powered by explosive cartridges 302 that are loaded in a magazine 304 and automatically fired under computer control in a breach 306. Explosive gases are routed through a check valve 308 to a pressurized-gas tank 310. A single 4-gang solenoid valve 312 and 314 steers high pressure gas to and vents gases from pressurized water tanks 316 and 318. When one tank is being pressured, the other is being vented. A high-water float control inlet valve 320 automatically admits water to pressurized water tank 316 when the liquid level is below the operating range maximum and the other tank 318 is receiving gas pressure from explosive-gas tank 310 through 4-gang solenoid valve 312. Another high-water float control inlet valve 322 admits water to pressurized water tank 318 when its liquid level is below its operating range maximum and its gases are vented. Similarly, a low-water float control inlet valve 324 shuts off water from pressurized water tank 316 when the liquid level falls below the operating range minimum. Another low-water float control outlet valve 326 shuts off water from pressurized water tank 318 when its liquid level is below its operating range minimum. Pressure safety valves (PSV) 330, 331, and 332 release overpressures to protect the respective tanks from rupturing.

A water turbine 340 converts the hydraulic flow through it to a mechanical torque applied to a rotating driveshaft 342. A second water turbine 344 converts its hydraulic flow to additional mechanical torque that is also applied to rotating driveshaft 342. A liquid circuit 346 returns to pressurized water tank 316 through high-water float switch and valve 320. A DC electrical generator 348 converts the rotating mechanical torque to electrical power that is converted to AC by an inverter 350. Gears and pulleys in front of the generator may be used to adjust the speed and power input. Fill and drain valves are connected to the various tanks as appropriate. The system control signals may be supported on a computer network or conventional process control loops and can involve wireless connections.

A controller 352 operates the magazine 304 and breach 306, and valves 312 and 314 to coordinate their timing, such that gas pressure from the pressurized-gas tank 310 is alternately routed to each pressurized water tank 316 and 318 until the liquid inside is pushed out into the other. The inverter 350 provides load sensing signals to the controller 352. A throttle control 354 applied to control motors on valves 324 and 326 can be used to control the power output of turbines 340 and 344.

FIG. 4 represents a two-stage electrical generator embodiment of the present invention, and is referred to herein by the general reference numeral 400. Generator 400 uses explosive gases to pressurize water, and then uses two stages of pressurized hydraulics to spin electrical generators with hydraulic motors and turbines. A first Stage-1 uses explosive cartridges to produce hot gases that will pressurize a tank 402. Computer timing and valve control 404 steers the high pressure gas alternately to a first hydraulic pressure tank-A 406 and then to a second hydraulic pressure tank-B 408 according to their respective liquid levels. Water passing from the pressurized one of the tanks to the non-pressurized one is used to spin a hydraulic pump or water turbine 410. Vent gases recovered from hydraulic pressure tank-A 406 and tank-B 408 are captured by a second stage gas pressure tank 412.

The pressure loss in the gas pressures between the first Stage-1 and second Stage-2 is a function of the differential volumes of hydraulic pressure tank-A 406 and tank-B 408 as they cycle between their minimum and maximum water levels.

The second stage gas pressure tank 412 supplies gas to a computer timing and valve control 414 steers the high pressure gas alternately to a third hydraulic pressure tank-C 416 and then to a fourth hydraulic pressure tank-D 418 according to their respective liquid levels. Water passing from the pressurized one of these tanks to the non-pressurized one is used to spin a hydraulic pump or water turbine 420.

Both water turbines 410 and 420 can be geared to drive a single DC electric generator 422. The electrical power produced is temporarily stored in batteries 424, and that can smooth out any voltage variations that would other wise result as the turbines are cycled between the hydraulic pressure tanks. An inverter 426 converts the DC power to AC power, and a transformer 428 is used to produce various commercial voltages, e.g., 110 VAC, 220-VAC, and 440-VAC at 50/60 Hertz.

Although the present invention has been described in terms of the presently preferred embodiments, it is to be understood that the disclosure is not to be interpreted as limiting. Various alterations and modifications will no doubt become apparent to those skilled in the art after having read the above disclosure. Accordingly, it is intended that the appended claims be interpreted as covering all alterations and modifications as fall within the "true" spirit and scope of the invention.

* * * * *

US008215111B1

(12) **United States Patent**
Richey

(10) Patent No.: **US 8,215,111 B1**
(45) Date of Patent: **Jul. 10, 2012**

(54) **ELECTRICAL GENERATION FROM EXPLOSIVES**

(76) Inventor: **Robert J. Richey**, Campbell, CA (US)

(*) Notice: Subject to any disclaimer, the term of this patent is extended or adjusted under 35 U.S.C. 154(b) by 502 days.

(21) Appl. No.: **12/386,822**

(22) Filed: **Apr. 23, 2009**

Related U.S. Application Data

(60) Provisional application No. 61/100,915, filed on Sep. 29, 2008.

(51) Int. Cl.
F01B 29/00 (2006.01)
F16D 31/00 (2006.01)
F16D 39/00 (2006.01)
(52) U.S. Cl. 60/512; 60/325; 60/415
(58) Field of Classification Search 60/369, 60/484, 516, 632, 634, 675, 914
See application file for complete search history.

(56) **References Cited**

U.S. PATENT DOCUMENTS

178,925 A	* 6-1876	Hardy	60/634
3,611,723 A	* 10/1971	Theis	60/327
3,648,458 A	* 3/1972	McAlister	60/415
3,650,572 A	* 3/1972	McClure	
4,301,774 A	11/1981	Williams	
5,551,237 A	* 9/1996	Johnson	60/6418
5,647,734 A	* 7/1997	Milleron	417/380
5,713,202 A	* 2/1998	Johnson	60/325
5,865,088 A	* 2/1999	Petschakis P.	91/4 R
6,182,615 B1	* 2/2001	Kershaw	123/19
6,739,331 B1	* 5/2004	Kershaw	60/512
7,531,908 B2	* 5/2009	Fries et al.	290/1 R

FOREIGN PATENT DOCUMENTS

WO WO 01/31197 A1 3/2001

* cited by examiner

Primary Examiner — Thomas Denion
Assistant Examiner — Christopher Jetton

(57) **ABSTRACT**

An electrical generator converts the high blast pressures of explosives into useful electricity by capturing the explosive gases and using the high gas pressures to alternately push water hydraulically between two tanks and through water turbines connected to DC electric generators. Water expelled through a water turbine from one tank is used to fill the other tank. Batteries can be used to store the electrical energy generated, and inverters followed by transformers convert the DC electric from the turbine-generators to 110-VAC, 220-VAC, and 440-VAC. A microcomputer controller connected to various sensors and solenoid valves coordinates the timing and routing of the detonation of explosives, tank pressures, venting, valving, and load control.

9 Claims, 4 Drawing Sheets

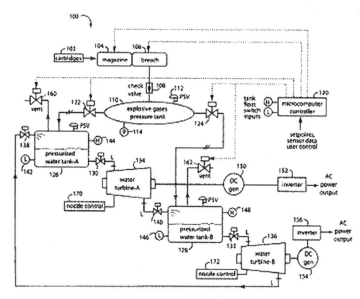

Fig. 1

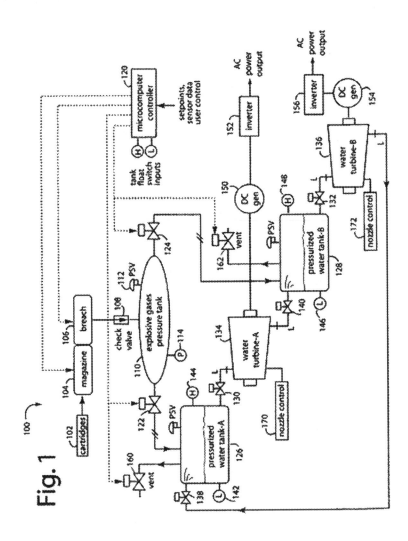

Fig. 2

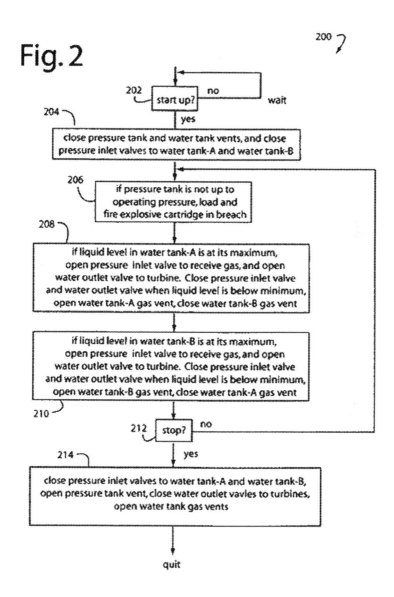

200

202 start up? — no → wait

204
close pressure tank and water tank vents, and close
pressure inlet valves to water tank-A and water tank-B

yes

206
if pressure tank is not up to
operating pressure, load and
fire explosive cartridge in breach

208
if liquid level in water tank-A is at its maximum,
open pressure inlet valve to receive gas, and open
water outlet valve to turbine. Close pressure inlet valve
and water outlet valve when liquid level is below minimum,
open water tank-A gas vent, close water tank-B gas vent

if liquid level in water tank-B is at its maximum,
open pressure inlet valve to receive gas, and open
water outlet valve to turbine. Close pressure inlet valve
and water outlet valve when liquid level is below minimum,
open water tank-B gas vent, close water tank-A gas vent

210

212 stop? — no

yes

214
close pressure inlet valves to water tank-A and water tank-B,
open pressure tank vent, close water outlet vavles to turbines,
open water tank gas vents

quit

Fig. 3

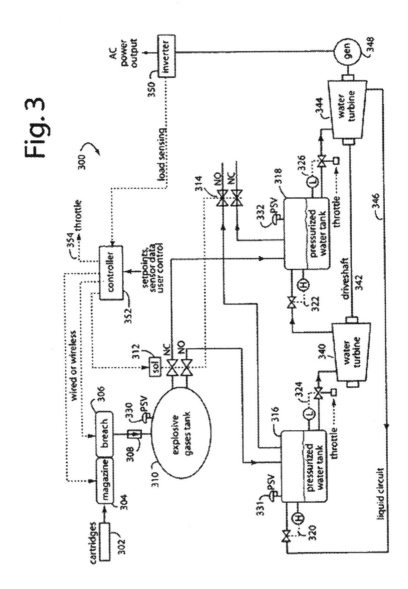

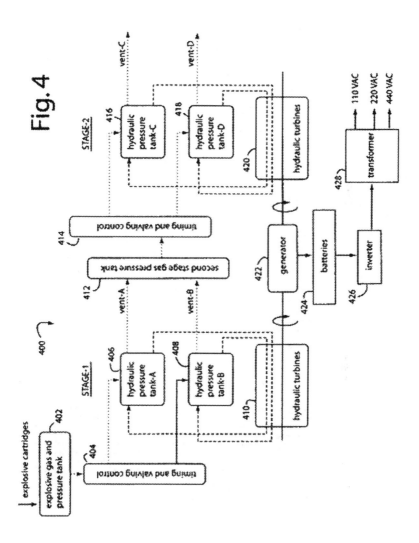

Fig. 4

1

ELECTRICAL GENERATION FROM EXPLOSIVES

RELATED APPLICATIONS

This Application claims benefit of U.S. Provisional Patent Application, A UNIQUE METHOD OF GENERATING ELECTRICITY USING EXPLOSIVE SUBSTANCES AS A POWER SOURCE, Ser. No. 61/100,915, filed Sep. 29, 2008.

BACKGROUND OF THE INVENTION

1. Field of the Invention

The present invention relates to electrical power generation, and in particular to devices and methods for converting the gas pressure generated by explosives into electricity.

2. Description of the Prior Art

Useful electrical energy does not exist in nature and it must be converted from other available energy forms such as gasoline, diesel, coal, natural gas, geothermal, steam, hydro, solar, etc. Some of these energy sources are expensive, some are highly pollutant, some are difficult to convert efficiently, and some are not very portable as is needed in vehicles.

Explosives like gunpowder; Cordite, Ballistite, and Poudre-B smokeless powders; trinitrotoluene (TNT), Dynamite, nitroglycerin; Tovex and other water gel explosives; etc., release a lot of energy in a very rapid pulse. Explosives usually have less potential energy than petroleum fuels, but their high rate of energy release produces large blast pressures. TNT has a detonation velocity of 6,940 m/s compared to 1,680 m/s for the detonation of a pentane-air mixture, and the 0.34-m/s stoichiometric flame speed of gasoline combustion in air. Explosives are classified as "low" or "high" explosives according to their rates of decomposition. Gunpowder is a low explosive, while TNT is a high explosive. Low explosives burn rapidly or deflagrate, while high explosives detonate.

The energy released includes high levels of heat, light, and gas pressure. These are all quickly dissipated if not captured or otherwise contained. For example, at 15° C. the volume of gas produced by the explosive decomposition of one mole of nitroglycerin, becomes, $V=(23.64 \text{ liter/mol})(7.25 \text{ mol})=171.4$ liters. The molar volume of an ideal gas at 15° C. is about 23.64 liters. The potential of an explosive is the total work that can be performed by the gas generated by the explosion. If uncontained, it expands adiabatically from its original volume until its pressure is reduced to atmospheric pressure and its temperature to ambient.

In the nitroglycerin reaction, $C_3H_5(NO_3)_3 \rightarrow 3CO_2 + 2.5H_2O + 1.5N_2 + 0.25O_2$, the products are carbon dioxide, water, nitrogen, oxygen, and heat. Therefore, a relatively small solid or liquid volume is converted into a very large volume of relatively benign gases. Nitroglycerin explosions are relatively clean, compared to TNT which is poisonous and produces a lot of carbon soot in its reaction.

Firearms and artillery use the gas pressure generated by the detonation of smokeless powder to accelerate bullets and projectiles to very high muzzle velocities on the order of 2,000+ feet per second. Sticks of explosives are detonated in holes drilled into geologic deposits to fracture the ores and make removing the material as easy as scooping up the pieces.

What is needed is a device and method to convert explosive energy into a more useful form of electrical energy as used in homes and industry.

SUMMARY OF THE INVENTION

Briefly, an electrical generator embodiment of the present invention converts the high blast pressures of explosives into

2

useful electricity by capturing the explosive gases and using the high gas pressures to alternately push water hydraulically between two tanks and through water turbines connected to DC electric generators. Water expelled through a water turbine from one tank is used to fill the other tank. Batteries can be used to store the electrical energy generated, and inverters followed by transformers convert the DC electric from the turbine-generators to 110-VAC, 220-VAC, and 440-VAC. A microcomputer controller connected to various sensors and solenoid valves coordinates the detonation of the explosives, tank pressures, venting, valving, and load control.

These and other objects and advantages of the present invention will no doubt become obvious to those of ordinary skill in the art after having read the following detailed description of the preferred embodiments which are illustrated in the various drawing figures.

IN THE DRAWINGS

FIG. 1 is a simplified functional block diagram of a single-stage electrical generator embodiment of the present invention that cycles pressurized water between two tanks and through two sets of water turbines;

FIG. 2 is a flowchart diagram of an electrical generator method embodiment of the present invention to cycle pressurized water between two tanks and two water turbines, as in FIG. 1;

FIG. 3 is a functional block diagram of a single-stage electrical generator embodiment of the present invention that cycles pressurized water between two tanks and through two water turbines like that of FIG. 1, but that reduces duplication of the DC generators and inverters, and some of the valving; and

FIG. 4 is a simplified functional block diagram of a two-stage electrical generator embodiment of the present invention that uses explosive gases to pressurize water, and then uses pressurized hydraulics to spin electrical generators with hydraulic motors and turbines.

DETAILED DESCRIPTION OF THE PREFERRED EMBODIMENT

In FIGS. 1-4 and the following text, some of the more conventional and routine elements commonly used with gas and hydraulic valves, pressure tanks, plumbing, and process control systems are not shown or described. For example, inspection ports and drains for water tanks, safety relief valves, check valves, nozzles for turbines, gearboxes and pulleys, wireless interfaces, wiring, etc. The components like these that should be used are engineering choices and are routinely stocked and installed by technicians. The critical and unusual combinations and their interrelationships are described here in detail.

FIG. 1 represents a single-stage electrical generator system embodiment of the present invention, and is referred to herein by the general reference numeral 100. Generator system 100 produces electrical power suitable for homes, businesses, industry, and the utility grid from the explosive energy captured from cartridges 102 loaded in a magazine 104 and fired in a breach 106. Cartridges 102 should include low explosives that burn clean and soot-free, and the chemical reactions should not produce any dangerous gases or byproducts. For example, nitroglycerin reactions only produce carbon dioxide, water, nitrogen, oxygen, and heat. The heat actually helps increase the gas pressures up to operating levels and should not be wasted or exhausted until the maximum in work has been extracted.

3

The heated gaseous explosive products are passed through a check valve 108 to a pressurized-gas tank 110. A pressure safety valve (PSV) 112 provides relief if the internal pressures exceed a safe maximum. A pressure sensor (P) 114 measures the tank pressures for a microcomputer controller 120. In some installations the pressure readings will be reported wirelessly, in others a simple 4-20 milliamp process control loop can be used.

Microcomputer controller 120 coordinates all the timing and valve control needed to operate generator system 100 and keep it safe. It uses readings from pressure sensor (P) 114 to determine when more cartridges 102 need to be loaded in magazine 104 and fired in breach 106, and it controls the actual firing. Microcomputer controller 120 also decides when and which gas pressure inlet valve 122 and 124 should be opened and closed for pressurized water tank-A 126 and pressurized water tank-B 128.

Pressurized water tank-A 126 and pressurized water tank-B 128 are not simultaneously pressurized, the pressure applied to them is alternated by gas pressure inlet valves 122 and 124 under control of microcomputer controller 120. What's important to the timing is the water levels inside the tanks, there are minimum and maximum operating levels that must be respected. Water inside one tank needs to flow out into the other tank through a water turbine, and the water cannot flow if the receiving water tank is pressurized at the same time.

In FIG. 1, an outlet valve-A 130 is opened to pass pressured liquid water (L) to a water turbine-A 134. Similarly, an outlet valve-B 132 is opened to pass pressured liquid water (L) to a water turbine-B 136. The liquid water returns from water turbine-B 136 through an inlet valve-A 138 back to water tank-A 126. Liquid water from water turbine-A 134 passes on through to inlet valve-B 140 to water tank-B 128.

The minimum and maximum operating levels of water that circulate between water tank-A 126 and water tank-B 128 are set by float switches (L, H) 142 and 144 for water tank-A 126, and by float switches (L, H) 146 and 148 for water tank-B 128. These float switches are connected to microcomputer controller 120, and the readings are used to determine when to open and close outlet valve-A 130, outlet valve-B 132, inlet valve-A 138, and inlet valve-B 140. The float switch connection could be done wirelessly, and a local loop could be included to automatically close, for example, water outlet valve-A 130 when minimum level float switch 142 senses low water.

Each water tank-A 126 and water tank-B 128 should be equipped with a water to add make up water, and to drain water completely, e.g., during maintenance.

In FIG. 1, water turbine-A 134 is mechanically connected by a rotating shaft to drive a DC electrical generator 150. The DC electrical power produced could be stored in batteries, and it is converted to AC electrical power by an inverter 152. Similarly, water turbine-B 136 is mechanically connected by a rotating shaft to drive another DC electrical generator 154. The DC electrical power produced could be stored in the same batteries, and it can also be independently converted to AC electrical power by an inverter 156. The voltage outputs of inverters 152 and 156 can be stepped-up or stepped-down by conventional transformers as needed, e.g., to 110-VAC, 220-VAC, and 440-VAC.

Microcomputer controller 120 is connected to sense the electrical loads placed on inverters 152 and 156, and uses the information to control how much pressurized water is needed to be passed through water turbine-A 134 and water turbine-B 136 to keep the overall operation in balance.

4

Once the pressurized gas inside the water tanks has done its job pushing out the water down to its minimum operating level, the residual pressurized gas can be vented out. For water tank-A 126, a vent valve-A 160 is used, and for water tank-B 128, a vent valve-A 162 is used. The residual gas pressures can be high enough to do useful work in a second stage generator. But any back pressure caused by the use of later stages can reduce the efficiency of the earlier stages by reducing the differential pressures between the pressurized tank and the vented one.

In operation, falling water levels inside the water tanks can be used by the minimum-level float switches 142 and 146 to trigger closed the associated water outlet valves 130 and 132. This, in turn can be used to trigger closed the gas pressure inlet valves 122 and 124, and to trigger open the gas pressure vent valves 160 and 162. Similarly, the maximum-level float switches 144 and 148 can be used to trigger closed the water inlet valves 138 and 140.

Pressurized water tank-A 126 and pressurized water tank-B 128 would normally be equipped with various conventional items not shown in FIG. 1. For example, inspection ports, drain valves, pressure gauges, pressure safety valves to release excess pressure, and a water make-up input to replace lost water.

Microcomputer controller 120 can increase and decrease the torque outputs of water turbine-A 134 and water turbine-B 136 by sending modulation controls to nozzle controls 170 and 172. Alternatively, water outlet valve-A 130 and water outlet valve-B 132 could be continuously adjustable, instead of simple fully open, fully closed solenoid types. Such torque modulation would be necessary in some applications to balance the power being generated with the loads applied. In such case, inverters 152 and 156 would also be required to provide load measurements to microcomputer controller 120.

FIG. 2 represents an electrical generator method embodiment of the present invention to cycle pressurized water between two tanks and two water turbines, as in FIG. 1. Such method is referred to herein by the general reference numeral 200. Method 200 is implemented as a computer program in software or firmware executed by a conventional microcomputer, e.g., microcomputer controller 120 (FIG. 1). Data inputs from sensors and switches are digitized for processing, user inputs are used to make process control decisions, and outputs to electro-mechanical solenoids are used to operate gas and hydraulic valves.

Method 200 includes three phases of operation: (1) startup, (2) power generation, and (3) shutdown. During startup, the operational pressures and valve settings must be initialized. During power generation, the gas pressure generated by the explosive cartridges must be switched between the two water tanks according to the respective water levels inside each. The amount of water forced between the water tanks and through the water turbines must be balanced with the electrical loads being placed on the system. During shutdown, the cycling must be stopped and the pressures relieved by opening the various vents.

Specifically, method 200 includes a step 202 for checking to see if the user wants to begin operation. If so, a step 204 closes the pressure tank and water tanks vents, and closes the inlet valves to the water tanks. A step 206 gets the gas pressure in the pressure tank up to operating levels by firing explosive cartridges as needed. A step 208 checks the water level inside water tank-A and if it's at its maximum operating level, a hydraulic cycle can begin. The gas inlet valve-A is opened, the gas vent valve-A is closed, and the water outlet valve-A to the associated turbine-A is opened. The gas pressure let in will push the water out through the outlet valve-A. When the

5

water level reaches minimum, the outlet valve-A is closed. The gas inlet valve-A is closed, and the gas vent valve-A is opened. The water inlet valve-A is opened to receive water from water tank-B.

A step 210 checks the water level inside water tank-B and if it's at its maximum operating level, a hydraulic cycle can begin. The gas inlet valve-B is opened, the gas vent valve-B is closed, and the water outlet valve-B to the associated turbine-B is opened. The gas pressure will push the water out through the outlet valve-B. When the water level in water tank-B reaches minimum, the outlet valve-B is closed. The gas inlet valve-B is closed, and the gas vent valve-B is opened. The water inlet valve-B is opened to receive water from water tank-A.

If the user is requesting a stop of operations, a step 212 passes control to a step 214. Otherwise, the process repeats in a loop back to step 206. Step 214 closes the gas inlet pressure valves to water tank-A and tank-B, opens the vents, and closes the water outlet valves to the turbines. Residual gas pressures inside the pressurized tank may be let down if another use cycle is not expected immediately.

FIG. 3 illustrates a single-stage system 300 that eliminates some of the duplication of the major components appearing in FIG. 1. System 300 assumes that when the water level in a water tank is below minimum, e.g., as detected by a low-water float switch, the water outlet valve should be closed. Similarly, when the water level in a water tank is above maximum, e.g., as detected by a high-water float switch, the water inlet valve should be closed. The gas inlet valve to a water tank can only be open if the gas vent is closed. The gas inlet valve to the water tank must be closed if the gas vent is open. The mechanisms implemented to enforce such logic can be built with relay logic, software, IC logic gates, and mechanical interlocks.

System 300 is powered by explosive cartridges 302 that are loaded in a magazine 304 and automatically fired under computer control in a breach 306. Explosive gases are routed through a check valve 308 to a pressurized-gas tank 310. A single 4-gang solenoid valve 312 and 314 steers high pressure gas to and vents gases from pressurized water tanks 316 and 318. When one tank is being pressured, the other is being vented. A high-water float control inlet valve 320 automatically admits water to pressurized water tank 316 when the liquid level is below the operating range maximum and the other tank 318 is receiving gas pressure from explosive-gas tank 310 through 4-gang solenoid valve 312. Another high-water float control inlet valve 322 admits water to pressurized water tank 318 when its liquid level is below its operating range maximum and its gases are vented. Similarly, a low-water float control inlet valve 324 shuts off water from pressurized water tank 316 when the liquid level falls below the operating range minimum. Another low-water float control outlet valve 326 shuts off water from pressurized water tank 318 when its liquid level is below its operating range minimum. Pressure safety valves (PSV) 330, 331, and 332 release overpressures to protect the respective tanks from rupturing.

A water turbine 340 converts the hydraulic flow through it to a mechanical torque applied to a rotating driveshaft 342. A second water turbine 344 converts its hydraulic flow to additional mechanical torque that is also applied to rotating driveshaft 342. A liquid circuit 346 returns to pressurized water tank 316 through high-water float switch and valve 320. A DC electrical generator 348 converts the rotating mechanical torque to electrical power that is converted to AC by an inverter 350. Gears and pulleys in front of the generator may be used to adjust the speed and power input. Fill and drain valves are connected to the various tanks as appropriate. The system control signals may be supported on a computer network or conventional process control loops and can involve wireless connections.

6

A controller 352 operates the magazine 304 and breach 306, and valves 312 and 314 to coordinate their timing, such that gas pressure from the pressurized-gas tank 310 is alternately routed to each pressurized water tank 316 and 318 until the liquid inside is pushed out into the other. The inverter 350 provides load sensing signals to the controller 352. A throttle control 354 applied to control motors on valves 324 and 326 can be used to control the power output of turbines 340 and 344.

FIG. 4 represents a two-stage electrical generator embodiment of the present invention, and is referred to herein by the general reference numeral 400. Generator 400 uses explosive gases to pressurize water, and then uses two stages of pressurized hydraulics to spin electrical generators with hydraulic motors and turbines. A first Stage-1 uses explosive cartridges to produce hot gases that will pressurize a tank 402. Computer timing and valve control 404 steers the high pressure gas alternately to a first hydraulic pressure tank-A 406 and then to a second hydraulic pressure tank-B 408 according to their respective liquid levels. Water passing from the pressurized one of the tanks to the non-pressurized one is used to spin a hydraulic pump or water turbine 410. Vent gases recovered from hydraulic pressure tank-A 406 and tank-B 408 are captured by a second stage gas pressure tank 412.

The pressure loss in the gas pressures between the first Stage-1 and second Stage-2 is a function of the differential volumes of hydraulic pressure tank-A 406 and tank-B 408 as they cycle between their minimum and maximum water levels.

The second stage gas pressure tank 412 supplies gas to a computer timing and valve control 414 steers the high pressure gas alternately to a third hydraulic pressure tank-C 416 and then to a fourth hydraulic pressure tank-D 418 according to their respective liquid levels. Water passing from the pressurized one of these tanks to the non-pressurized one is used to spin a hydraulic pump or water turbine 420.

Both water turbines 410 and 420 can be geared to drive a single DC electric generator 422. The electrical power produced is temporarily stored in batteries 424, and that can smooth out any voltage variations that would otherwise result as the turbines are cycled between the hydraulic pressure tanks. An inverter 426 converts the DC power to AC power, and a transformer 428 is used to produce various commercial voltages, e.g., 110 VAC, 220-VAC, and 440-VAC at 50/60 Hertz.

Although the present invention has been described in terms of the presently preferred embodiments, it is to be understood that the disclosure is not to be interpreted as limiting. Various alterations and modifications will no doubt become apparent to those skilled in the art after having read the above disclosure. Accordingly, it is intended that the appended claims be interpreted as covering all alterations and modifications as fall within the "true" spirit and scope of the invention.

The invention claimed is:

1. A generator system, comprising:

a high pressure gas tank providing for the capture and confinement of gases generated by an explosive cartridge;

a magazine and breach connected to the high pressure gas tank, and providing for the operation of said explosive cartridge;

a pair of interconnected liquid tanks connected to receive gases routed from the high pressure gas tank, said liquid tanks containing a liquid and interconnected such that said liquids within flow in a circuit between the liquid tanks;

a high pressure turbine connected to be driven by said liquids flowing between the liquid tanks;

a low pressure gas tank connected to the pair of interconnected liquid tanks, and providing for the capture and confinement of gases vented from the interconnected liquid tanks;

a second pair of interconnected liquid tanks connected to receive gases routed from the low pressure gas tank, said liquid tanks containing a liquid and interconnected such that said liquids within flow in a circuit between the liquid tanks;

a low pressure turbine connected to be driven by said liquids flowing between the second pair of interconnected liquid tanks;

an electric generator connected to be driven by the high pressure and low pressure turbines and able to produce electrical power;

and a controller to operate valves and to coordinate the timing such that gas pressure from the pressurized gas tank is alternately routed to each liquid tank until the liquid inside is pushed out into the other;

wherein energy from said explosive cartridge is converted into electrical power.

2. The system of claim 1, further comprising: a high-water float switch and a low-water float switch disposed in each of the liquid tanks and connected to the controller; wherein the controller is enabled to maintain said liquid levels within each pair of interconnected liquid tanks over an operational range.

3. The system of claim 1, further comprising:

a liquid inlet valve providing a controlled input for each of the pair of interconnected liquid tanks that is connected in a circuit to receive liquids from the other liquid tank in the pair.

4. The system of claim 1, further comprising:

a liquid outlet valve providing a controlled output for each of the pair of interconnected liquid tanks that is connected in a circuit to transmit liquids to the other liquid tank in the pair.

5. The system of claim 1, further comprising:

a liquid inlet valve providing a controlled input for each of the pair of interconnected liquid tanks that is connected in a circuit to receive liquids from the other liquid tank in the pair;

a liquid outlet valve providing a controlled output for each of the pair of interconnected liquid tanks that is connected in a circuit to transmit liquids to the other liquid tank in the pair; and

a high-water float switch and a low-water float switch disposed in each of the liquid tanks;

wherein the liquid inlet valve and liquid outlet valve are controlled by the controller according to signals obtained from the high-water float switch and a low-water float switch.

6. A generator system, comprising:

a pressurized-gas tank providing for the capture and confinement of gases generated by an explosive cartridge;

a magazine and breach connected to the pressurized-gas tank, and providing for the operation of said explosive cartridge;

a pair of interconnected first and second liquid tanks connected to receive gases routed from the pressurized-gas tank, said liquid tanks containing a liquid and interconnected such that said liquids within flow in a circuit between the liquid tanks;

a first and a second liquid inlet valve providing a controlled input for each of the pair of interconnected liquid tanks that is connected in a circuit to receive liquids from the other liquid tank in the pair;

a first and a second liquid outlet valve providing a controlled output for each of the pair of interconnected liquid tanks that is connected in a circuit to transmit liquids to the other liquid tank in the pair;

a high-water float switch and a low-water float switch disposed in each of the first and second liquid tanks;

a first turbine connected to be driven by said liquids flowing from said first liquid tank to said second liquid tank;

a second turbine connected to be driven by said liquids flowing from said second liquid tank to said first liquid tank;

an electric generator connected to be driven by at least one of the first and second turbines and able to produce electrical power;

and a controller to operate the magazine and breach, and first and a second liquid inlet valve, and first and a second liquid outlet valves to coordinate their timing, such that gas pressure from the pressurized-gas tank is alternately routed to each liquid tank until the liquid inside is pushed out into the other, and according to signals obtained from the high-water float switch and low-water float switch;

wherein energy from said explosive cartridge is converted into electrical power.

7. A method of converting explosives energy into electrical power, comprising a computer program in software or firmware executed by a conventional microcomputer, with data inputs from sensors and switches digitized for processing, user inputs used to make process control decisions, and outputs to electro-mechanical solenoids to operate gas and hydraulic valves, comprising:

closing pressure tank and water tank vents, and closing water tank inlet valves;

raising a gas pressure in a pressure tank up to an operating level by firing explosive cartridges as needed;

checking a water level inside a water tank-A and if it's at its maximum operating level, beginning a hydraulic cycle, wherein a gas inlet valve-A is opened, a gas vent valve-A is closed, and a water outlet valve-A to an associated turbine-A is opened; wherein, a gas pressure let in can push water out through said outlet valve-A until the water level reaches a minimum, and said outlet valve-A is closed, said gas inlet valve-A is closed, and gas vent valve-A is opened, and water inlet valve-A is opened to receive water from water tank-B;

checking a water level inside a water tank-B and if it's at its maximum operating level, beginning a hydraulic cycle, wherein a gas inlet valve-B is opened, a gas vent valve-B is closed, and a water outlet valve-B to an associated turbine-B is opened; wherein, a gas pressure let in can push water out through said outlet valve-B until the water level reaches a minimum, and said outlet valve-B is closed, said gas inlet valve-B is closed, and gas vent valve-B is opened, and water inlet valve-B is opened to receive water from water tank-A.

8. The method of claim 7, further comprising: if a user is not requesting a stop of operations, the process repeats in a loop.

9. The method of claim 7, further comprising: if a user is requesting a stop of operations, closing gas inlet pressure valves to water tank-A and tank-B, opening gas vents, and closing said water outlet valves to the turbines; wherein, residual gas pressures inside said pressurized tank is let down if another use cycle is not scheduled immediately.

* * * * *

AN ALTERNATE EXPLOSIVE SUBSTANCE TO BE USED WITH PATENT US 8215111

The Substance: Petaerythritol Tetrannitrate

This Chemical is a nitrate ester of pentaerythritol. Penta refers to the five carbon atoms of the neopentane skeleton.

It is the most widely known explosive and is the most powerful explosive as well.

If mixed with a plasticizer it forms a plastic explosive. If mixed instead with RDX and other minor additives (not identified), it forms another plastic explosive called Semetex also. It has been in existence since 1891 and was synthesized by the nitration of pentaerythritol.

PETN, as it is commonly called, is practically insoluble in water. Due to its atomic structure it is also resistant to attacks by many chemical agents. This means it is very stable. Its stability allows it to be stored for extended periods of time without significant deterioration.

It has low volatility, low solubility, is medium to shock sensitivity and frictional sensitivity as well.

Production is accomplished by the reaction of pentaerythritol with concentrated nitric acid which forms a precipitate that can be re-crystallized from acetone to give processable crystals.

In commercial production it is manufactured by numerous manufacturers all over the world and is in large supply. Its consistency is similar to fine popcorn salt.

The most common use is as an explosive with high brisance. Brisance is rapidity of explosion or shattering rate of expansion which is undesirable in this case. There is difficulty in its detonation. This contributes to safe handling as dropping it does not usually result in an explosion. Also accidental contact may not be serious as trans-dermal absorption is low. Igniting it typically does not result in an explosion. There is difficulty in igniting it and its burn rate is relatively slow. This is an advantage as concerns using it as the explosive of choice for this Invention. It can be detonated by a laser or an electric spark of a specified intensity. It appears that it can be supplied in cartridge form.

The volume of expansion converting from a solid to a gas is 790 dm3 /kg.

Its molecular formula is C5H8N4O12 and exists as a white crystalline solid.

ABOUT THE AUTHOR

This author earlier in life was an aerospace engineer for six years. Three of those years were devoted to performing analysis on various components of the outer shells of the Apollo command modules. Three other years involved being in an environmental group on a project for the Douglas Space Center. Eleven years were spent as a registered professional engineer in traffic engineering in California while employed by a county in California. Other projects involved designing the large traffic signs on California State Highway 85 and computerizing several expressway-signalized interconnected systems as a consultant. This author is presently retired.

Printed in the United States
By Bookmasters